ENERGY

FACTS AND FIGURES

Robert H. Romer

Spring Street Press
104 Spring Street
Amherst
Massachusetts 01002

Published by

Spring Street Press
104 Spring Street
Amherst
Massachusetts
01002

(413) 253-7748
or
(413) 542-2258

Library of Congress Cataloging in Publication Data

Romer, Robert H.

Energy facts and figures.

1. Power resources--Statistics.   2. Power resources.   I. Title.

TJ163.2.R65   1984   333.79   84-16278

ISBN 0-931691-17-6

CONTENTS

INTRODUCTION

'Let us not underrate the value of a fact;
it will one day flower in a truth.'
- Henry David Thoreau, 1842

In publishing this collection of data, my major goal is to provide the essential information for understanding energy problems. Anyone with an interest in energy, anyone who sometimes wants to check up on a figure in a news report, will find this collection useful as a convenient but not overwhelming 'energy handbook'. [Much of this information originally appeared in the appendices to my book, Energy, An Introduction to Physics (W.H.Freeman, 1976), and is reproduced here with permission of the publisher.] The tables and graphs in Sections I and K, in particular, provide a 17-page energy history of the world and of the United States, from 1850 to the present. Section L provides a more detailed look at energy use in the United States in 1983, and Section P shows how energy prices paid by the typical consumer have changed over the years since 1960. Every effort has been made to provide accurate and consistent information, and to base every numerical value and every graph on sources that are as close to primary sources as possible. Although all of the information on which this energy history is based is available to the public, all of the tables and graphs are original.

In other sections of this work, I provide, among other things, physical constants, the dimensions of our solar system, information about the earth and its atmosphere, the energy content of various fuels, degree-day and solar energy data, energy requirements for electrical appliances and for various modes of passenger and freight transportation, and - in particular - an extensive and easy to use table of units and conversion factors (Section A).

Although many of the numbers here are identical to those in Energy, An Introduction to Physics, there is much new material in Sections I, K, L and P, which now contain data through 1983. A dramatic demonstration of the fact that some things really have changed in the past decade is provided by the simple exercise of comparing the graphs presented here in Sections I and K with those in my earlier book, where the data end in 1973, or simply by first looking at some of the present graphs with the most recent decade concealed and then looking at the full graph. With few exceptions, the pre-1973 data show nothing but apparently perpetual growth; with similarly few exceptions, almost every graph in Sections I and K now shows strikingly different behavior during the past ten years. (Not unrelated to this phenomenon is a number in Section N that has been left as it was but should really now be revised: 12 miles/gallon is no longer a fair figure for the gasoline mileage of the 'typical American automobile'.) However, preliminary data for 1984 suggest at least a temporary reversal of recent trends. United States energy consumption was probably substantially higher in 1984 than in 1983, up from 71 mQ to about 77 mQ. (1 mQ = 1 quad = 10^{15} BTU; see below.) The percentage of our energy that we import also appears to have increased, from 12% in 1983 to about 15% in 1984.

A few other numbers should be very slightly revised but have been allowed to stand. The value in Figure E.3 for the rate of decrease of fossil fuel energy, as well as that for the rate of conversion of energy by human activities into thermal form, should be increased to about 10^{13} watts (0.006% of the solar input), while that for the rate of decrease of nuclear energy should be increased to 3×10^{11} watts (0.0002%). All three energy conversion rates are still insignificant in comparison with the solar input, but associated processes such as CO_2 production may be having significant effects on world climate.

Account has been taken of the use of wind, geothermal and solar energy, and the combustion of wood, refuse, etc. for electrical generation. No attempt has been made, however, to present numerical information on the use of energy from these sources for purposes other than electrical generation. The one exception to this statement is the inclusion of estimated consumption of wood for all energy purposes in Figure K.6, in which the entire energy history of the United States is displayed on a single semilogarithmic graph.

In my earlier book I relied heavily on the 1972 Stanford Research Institute report, 'Patterns of Energy Consumption', especially for information on how energy was used within the residential, commercial, industrial and transportation sectors of the economy. In the absence of comparable more recent data, Tables L.7, 8 and 9 show the 1973 distributions, giving <u>percentages</u> of sector and national totals. Perhaps a more thorough revision will appear in a future edition of <u>Energy</u> <u>Facts</u> <u>and</u> <u>Figures</u>.

<u>Hydroelectric power</u>. I have made one important change in my 'energy bookkeeping': specifically, in the treatment of hydroelectric energy, in calculating the 'energy consumption' for hydroelectric generation. In <u>Energy</u>, <u>An</u> <u>Introduction</u> <u>to</u> <u>Physics</u>, I treated hydroelectric generation as if its efficiency were 100%, a fair approximation to the true efficiency of 85% or so, and far higher than the efficiencies of about 30% that characterize steam-electric power plants. For example, the amount of electrical energy generated from hydropower in 1970 in the United States was 247×10^9 kWh (Table K.4B), or, after this amount of energy is converted to quads or mQ (see below), the electrical energy generated at hydroelectric plants was 0.84 mQ. This figure was therefore entered in Table K.2B as the 'energy consumption' for hydroelectric generation. In thus using identical values for the electrical energy <u>generated</u> from hydropower and the energy <u>consumed</u> in hydroelectric generation, I was following the procedure used in 'Patterns of Energy Consumption'. Although this practice has a logical appeal to a physicist like myself, it is rarely used in tabulations of energy data. The usual procedure, the one I have now adopted, is to treat hydroelectric generation as if its efficiency were equal to that of the average efficiency of fossil-fuel plants. Thus for 1970 we again have 0.84 mQ of energy generated from hydroelectric power, but now 2.60 mQ, about three times larger, is the 'energy <u>consumed</u> in hydroelectric generation'. Therefore we also have 1.76 mQ of 'waste heat'. Although there is no physical basis for the invention of this fictitious waste heat (whose inclusion artificially inflates the United States total energy consumption for 1970 from 65.4 mQ to 67.1 mQ), there is a rational argument in favor of this procedure: it makes possible an easier comparison of the effect of substituting, say, a hydroelectric generating plant with an output of 1 GW for a fossil-fuel plant with the same output capacity. In any event, I have now fallen into line and abandoned the procedure used in 'Patterns of Energy Consumption' in favor of the more common one.

<u>Q, mQ, and quad</u>. In my 1976 book, I adopted the Q, defined as 10^{18} BTU, an energy unit in use since at least the early 1950s. I also needed a name for 10^{15} BTU, and I made the natural choice: 1 milli-Q (mQ) = 10^{15} BTU. At about the time I was finishing my book, the term 'quad', also equal to 10^{15} BTU (1 'quadrillion' BTU) began to appear in writings about energy problems. I have, for the sake of simplicity, retained the milli-Q, but the translation is easy:

$$1 \text{ mQ} = 1 \text{ quad} = 10^{15} \text{ BTU}.$$

(Annual energy consumption in the United States is about 70 quads = 70 mQ = 0.07 Q.) Interchangeable use of mQ and quad would cause no difficulty were it not for the fact that a few authors, presumably ignorant of the prior history of the name, have used Q as an abbreviation for the quad! Fortunately, this practice is not common. It is also fortunate that an apparent discrepancy of a factor of 1000 is so large that even in discussions of world-wide energy resources, where order of magnitude estimates are often necessary and desirable, a mistake this large is not likely to go unnoticed.

<u>Petroleum, Oil, NGLs, Natural Gas, Coal, Lignite.</u> In most places, I use 'oil' to include both crude oil as obtained from oil wells (and the refined products made from it) and 'natural gas liquids' (NGLs), which are fuels such as propane and butane obtained in the processing of natural gas. That is, 'oil' refers to the total amount of petroleum <u>liquids</u>. In some of the tables and graphs in Section I, however, crude oil and NGL production are treated separately. Figures for natural gas refer to 'dry' natural gas, the amount remaining after the NGLs have been removed. (During most of our history, much of the natural gas found in foreign oil fields has been wasted by venting or burning; production of natural gas and NGLs outside the United States has therefore been low.) 'Coal' includes bituminous coal (the major component), anthracite coal, and lignite. Lignite (sometimes called 'brown coal') is a coal with a somewhat smaller fuel value than other coals. Lignite has not been of much importance in the United States in the past, but a significant part of the coal in the western United States is in this form; in coming decades, lignite will probably account for a larger fraction of our coal production.

<u>Units.</u> Some purists have urged me to use nothing but approved SI units (Système International). This, I think, would be a mistake. I believe there are good reasons for the existence of a variety of units. There is also something appealing about the existence of many kinds of units. Perhaps it would be logically preferable to have just one set of units, but until that day arrives, we might as well take some pleasure in the diversity of the units we have inherited. Furthermore, were some of us to forsake the kWh, the BTU and the mQ (or quad), we would create unnecessary obstacles to communication with many of those who are most actively involved with the energy problem. There are certainly instances, though, in which idiosyncratic units serve to conceal truth and not to reveal it. One is in tabulations of the energy content of various fuels (similar to that in Section H), where it is common to give the fuel value of natural gas in BTU/ft^3, that of wood in BTU/cord, and so on. To me it was a revelation when I first calculated all of these fuel values in terms of energy per unit mass, when I immediately saw that they are all of nearly the same size. All of the fuel values in Section H (except, of course, those for fission and fusion) are, to within an order of magnitude, 10^7 J/kg. That figure is really about 0.1 electron-volts per atomic mass unit. That is, combustion of any fuel is (of course!) a chemical reaction involving an energy release of about an electron-volt, in the reaction of atoms with a mass of the order of 10 atomic mass units.

Some of the most valuable units, incidentally, are the 'informal' ones: 1 barrel of oil = 5.6×10^6 BTU, 1 jelly donut = 10^6 joules, 1 (big) power plant = 1 GW (1 gigawatt, 10^9 watts), 1 person = 1 light bulb = 100 W. (Averaged over all hours of day and night, the rate at which a typical person consumes food energy - which is then nearly all converted to heat - is about 100 watts.) In 1983, the installed generating capacity of the United States was about 700 'power plants'. How much more meaningful this is than the correct, but sterile, figure of 7×10^{11} watts. (And how much easier it is to catch a gross error. It takes some reflection before one realizes that a figure of, say, 7×10^{14} watts is wrong, but if one were told that 'the generating capacity of the United States is 700,000 large power plants', one would immediately see that <u>some</u> mistake had been made.)

But one must have a simple and foolproof method for correctly converting units. Here is one method that can be applied with an absolute minimum of thought. You can always multiply a quantity by <u>unity</u> (as many times as you like) without changing its real value, and unity can be written in various forms: 1 = 1 mile/5280 ft = 60 seconds/1 minute, etc. Here is one simple example. The flow rate of water over Niagara Falls, given as 1.2×10^8 gallons/minute, is to be converted to m^3/second:

$$1.2 \times 10^8 \; \frac{\text{gallons}}{\text{minute}} \times \frac{1 \text{ minute}}{60 \text{ seconds}} \times \frac{3.785 \times 10^{-3} \text{ m}^3}{1 \text{ gallon}} = 7570 \text{ m}^3/\text{second}.$$

You need nothing but a conversion table, such as that supplied in Section A. It is almost impossible to make a mistake. The worst that can happen is that one of the factors of unity is put in 'upside down', perhaps leading to the strange-looking result 5.28×10^8 gallons^2/m^3-sec, a result that is not really <u>wrong</u> but is obviously not in the desired form.

A few physicists have asked me to delete the pound from the mass table, and the kilogram from the force table (in spite of the qualifying remarks that appear in those tables). Just as I find it desirable to keep the BTU and the kWh, I (like most people) can find meaning in the statement that 1 kg = 2.2 lbs, though as a physicist I often wish to attach several paragraphs of explanation to that particular equality sign.

<u>Significant figures and accuracy.</u> Although the conversion factors in Section A are accurate to four figures, the same cannot be said of some of the other numbers in this book. I have not tried to be completely consistent in rounding off data. I give, for instance, United States energy consumption for 1983 as 70.6 mQ or 71 mQ (in Sections K and L), probably a reasonable number of figures, but readers may do their own rounding in other cases. No one, for instance, will believe that average per capita power consumption in the United States in 1983 was 10074 watts (to five figures), or that the 'equivalent gasoline mileage' of a bicyclist (Section N) is 1560 miles/gallon (to three or four figures). Other numbers, such as estimates of eventual total world-wide coal and oil production, are subject to much greater uncertainties; debate on the numerical values, even on definitions of terms, continues to provide full-time occupation to many people.

<u>Sources.</u> The major sources used for my original compilation of energy data were listed in my earlier book. The following serial publications are those I now find most helpful in providing current information.
Annual Energy Review (Energy Information Administration)
Energy Statistics Yearbook (United Nations)
International Energy Annual (Energy Information Administration)
Monthly Energy Review (Energy Information Administration)
Statistical Abstract of the United States (Bureau of the Census)
Statistical Yearbook of the Electric Utility Industry (Edison Electric Institute)

<u>Acknowledgments.</u> I am indebted to Theodore Romer, Betty Steele and George Watson for frequent assistance in diagnosing malfunctioning computer programs, to Margaret Groesbeck and Floyd Merritt of the Amherst College library for help in obtaining reference materials, and to Irene Haynes, Christina Romer, David Romer, Diana Romer, Evan Romer and Ruth Romer for abundant help, advice, and criticism. I also want to thank my physicist colleagues at Amherst College and the many readers who have expressed admiration for the appendices to <u>Energy - An Introduction to Physics</u> and have urged me to bring them up to date.

Robert H. Romer
Amherst, Massachusetts
October, 1984

A
Units and Conversion Factors

TABLE A.1 Area.

1 square centimeter (cm²) =	1 square inch (in.²) =	1 square foot (ft²) =	1 square meter (m²) =
1 cm²	6.452 cm²	929 cm²	10^4 cm²
0.155 in.²	1 in.²	144 in.²	1550 in.²
1.076×10^{-3} ft²	6.944×10^{-3} ft²	1 ft²	10.76 ft²
10^{-4} m²	6.452×10^{-4} m²	0.0929 m²	1 m²
2.471×10^{-8} acre	1.594×10^{-7} acre	2.296×10^{-5} acre	2.471×10^{-4} acre
10^{-8} hectare	6.452×10^{-8} hectare	9.29×10^{-6} hectare	10^{-4} hectare
10^{-10} km²	6.452×10^{-10} km²	9.29×10^{-8} km²	10^{-6} km²
3.861×10^{-11} mile²	2.491×10^{-10} mile²	3.587×10^{-8} mile²	3.861×10^{-7} mile²

1 acre =	1 hectare (ha) =	1 square kilometer (km²) =	1 square mile (mile²) =
4.047×10^7 cm²	10^8 cm²	10^{10} cm²	2.59×10^{10} cm²
6.273×10^6 in.²	1.55×10^7 in.²	1.55×10^9 in.²	4.014×10^9 in.²
4.356×10^4 ft²	1.076×10^5 ft²	1.076×10^7 ft²	2.788×10^7 ft²
4047 m²	10^4 m²	10^6 m²	2.59×10^6 m²
1 acre	2.471 acres	247.1 acres	640 acres
0.4047 hectare	1 hectare	100 hectares	259 hectares
4.047×10^{-3} km²	0.01 km²	1 km²	2.59 km²
1.562×10^{-3} mile²	3.861×10^{-3} mile²	0.3861 mile²	1 mile²

TABLE A.2 Density.

1 kilogram per cubic meter (kg/m³) =	1 pound per cubic foot (lb/ft³) =
1 kg/m³	16.02 kg/m³
6.243×10^{-2} lb/ft³	1 lb/ft³
8.345×10^{-3} lb/gal	0.1337 lb/gal
0.001 g/cm³	1.602×10^{-2} g/cm³
3.613×10^{-5} lb/in.³	5.787×10^{-4} lb/in.³

1 pound per gallon (lb/gal) =	1 gram per cubic centimeter* (g/cm³) =	1 pound per cubic inch (lb/in.³) =
119.8 kg/m³	1000 kg/m³	2.768×10^4 kg/m³
7.481 lb/ft³	62.43 lb/ft³	1728 lb/ft³
1 lb/gal	8.345 lb/gal	231 lb/gal
0.1198 g/cm³	1 g/cm³	27.68 g/cm³
4.329×10^{-3} lb/in.³	3.613×10^{-2} lb/in.³	1 lb/in.³

*Note that since the density of water is almost precisely 1 g/cm³, these data give the density of water as 1000 kg/m³, 62.43 lb/ft³, etc.

TABLE A.3 Energy.

1 electron-volt (eV) =	1 million electron-volts (MeV) =	1 joule (J) =	1 calorie (cal) =
1 eV	10^6 eV	6.241×10^{18} eV	2.611×10^{19} eV
10^{-6} MeV	1 MeV	6.241×10^{12} MeV	2.611×10^{13} MeV
1.602×10^{-19} J	1.602×10^{-13} J	1 J	4.184 J
3.829×10^{-20} cal	3.829×10^{-14} cal	0.239 cal	1 cal
1.52×10^{-22} BTU	1.52×10^{-16} BTU	9.485×10^{-4} BTU	3.968×10^{-3} BTU
3.829×10^{-23} kcal	3.829×10^{-17} kcal	2.39×10^{-4} kcal	0.001 kcal
4.451×10^{-26} kWh	4.451×10^{-20} kWh	2.778×10^{-7} kWh	1.162×10^{-6} kWh
1.52×10^{-28} MBTU	1.52×10^{-22} MBTU	9.485×10^{-10} MBTU	3.968×10^{-9} MBTU
1.854×10^{-30} MW-day	1.854×10^{-24} MW-day	1.157×10^{-11} MW-day	4.843×10^{-11} MW-day
5.077×10^{-33} MW-yr	5.077×10^{-27} MW-yr	3.169×10^{-14} MW-yr	1.326×10^{-13} MW-yr
1.52×10^{-37} mQ	1.52×10^{-31} mQ	9.485×10^{-19} mQ	3.968×10^{-18} mQ
1.52×10^{-40} Q	1.52×10^{-34} Q	9.485×10^{-22} Q	3.968×10^{-21} Q

1 British Thermal Unit (BTU) =	1 kilocalorie (kcal or Cal) =	1 kilowatt-hour (kWh) =	1 million BTU (MBTU) =
6.581×10^{21} eV	2.611×10^{22} eV	2.247×10^{25} eV	6.581×10^{27} eV
6.581×10^{15} MeV	2.611×10^{16} MeV	2.247×10^{19} MeV	6.581×10^{21} MeV
1054 J	4184 J	3.6×10^6 J	1.054×10^9 J
252 cal	1000 cal	8.604×10^5 cal	2.52×10^8 cal
1 BTU	3.968 BTU	3413 BTU	10^6 BTU
0.252 kcal	1 kcal	860.4 kcal	2.52×10^5 kcal
2.929×10^{-4} kWh	1.162×10^{-3} kWh	1 kWh	292.9 kWh
10^{-6} MBTU	3.968×10^{-6} MBTU	3.413×10^{-3} MBTU	1 MBTU
1.22×10^{-8} MW-day	4.843×10^{-8} MW-day	4.167×10^{-5} MW-day	0.0122 MW-day
3.341×10^{-11} MW-yr	1.326×10^{-10} MW-yr	1.141×10^{-7} MW-yr	3.341×10^{-5} MW-yr
10^{-15} mQ	3.968×10^{-15} mQ	3.413×10^{-12} mQ	10^{-9} mQ
10^{-18} Q	3.968×10^{-18} Q	3.413×10^{-15} Q	10^{-12} Q

1 megawatt-day (MW-day) =	1 megawatt-year (MW-yr) =	1 milli-Q (mQ) =	1 Q =
5.393×10^{29} eV	1.97×10^{32} eV	6.581×10^{36} eV	6.581×10^{39} eV
5.393×10^{23} MeV	1.97×10^{26} MeV	6.581×10^{30} MeV	6.581×10^{33} MeV
8.64×10^{10} J	3.156×10^{13} J	1.054×10^{18} J	1.054×10^{21} J
2.065×10^{10} cal	7.542×10^{12} cal	2.52×10^{17} cal	2.52×10^{20} cal
8.195×10^7 BTU	2.993×10^{10} BTU	10^{15} BTU	10^{18} BTU
2.065×10^7 kcal	7.542×10^9 kcal	2.52×10^{14} kcal	2.52×10^{17} kcal
2.4×10^4 kWh	8.766×10^6 kWh	2.929×10^{11} kWh	2.929×10^{14} kWh
81.95 MBTU	2.993×10^4 MBTU	10^9 MBTU	10^{12} MBTU
1 MW-day	365.2 MW-day	1.22×10^7 MW-day	1.22×10^{10} MW-day
2.738×10^{-3} MW-yr	1 MW-yr	3.341×10^4 MW-yr	3.341×10^7 MW-yr
8.195×10^{-8} mQ	2.993×10^{-5} mQ	1 mQ	1000 mQ
8.195×10^{-11} Q	2.993×10^{-8} Q	0.001 Q	1 Q

Other units of energy:

1 erg $= 10^{-7}$ J

1 foot-pound (ft-lb) $= 1.356$ J

1 therm $= 10^5$ BTU, often used in reporting sales of natural gas. (1 therm is almost exactly the energy content of 100 cubic feet of natural gas.)

1 kiloton $= 10^{12}$ cal $= 4.184 \times 10^{12}$ J, approximately the energy released in the explosion of 1000 tons of TNT. The kiloton (and the megaton) are frequently used in referring to A-bomb and H-bomb explosions.

1 horsepower-hour (hp-hr) $= 0.746$ kWh $= 2.686 \times 10^6$ J, the energy delivered by 1 hp acting for 1 hour.

TABLE A.4 Fluid flow rate.

1 gallon per day =	1 acre-foot per year =	1 cubic foot per minute (ft³/min or cfm) =
1 gal/day	892.2 gal/day	1.077×10^4 gal/day
1.121×10^{-3} acre-ft/yr	1 acre-ft/yr	12.07 acre-ft/yr
9.283×10^{-5} ft³/min	8.282×10^{-2} ft³/min	1 ft³/min
1.547×10^{-6} ft³/sec	1.38×10^{-3} ft³/sec	1.667×10^{-2} ft³/sec
4.381×10^{-8} m³/sec	3.909×10^{-5} m³/sec	4.719×10^{-4} m³/sec
10^{-9} bgd	8.922×10^{-7} bgd	1.077×10^{-5} bgd

1 cubic foot per second (ft³/sec or cfs) =	1 cubic meter per second (m³/sec) =	1 billion gallons per day (bgd) =
6.463×10^5 gal/day	2.282×10^7 gal/day	10^9 gal/day
724.4 acre-ft/yr	2.558×10^4 acre-ft/yr	1.121×10^6 acre-ft/yr
60 ft³/min	2119 ft³/min	9.283×10^4 ft³/min
1 ft³/sec	35.31 ft³/sec	1547 ft³/sec
2.832×10^{-2} m³/sec	1 m³/sec	43.81 m³/sec
6.463×10^{-4} bgd	2.282×10^{-2} bgd	1 bgd

TABLE A.5 Force.

1 gram (g) =	1 newton (N) =	1 pound (lb) =	1 kilogram (kg) =	1 ton =
1 g	102 g	453.6 g	1000 g	9.072×10^5 g
9.807×10^{-3} N	1 N	4.448 N	9.807 N	8896 N
2.205×10^{-3} lb	0.2248 lb	1 lb	2.205 lb	2000 lb
0.001 kg	0.102 kg	0.4536 kg	1 kg	907.2 kg
1.102×10^{-6} ton	1.124×10^{-4} ton	0.0005 ton	1.102×10^{-3} ton	1 ton

Other units of force: 1 dyne = 10^{-5} N; 1 ounce = 1/16 lb = 0.278 N.

Note. The kilogram and the gram are, strictly speaking, units of mass, not of force. A "force of 1 kg" is the gravitational force exerted by the earth on a mass of 1 kg, where the acceleration due to gravity has its standard value.

TABLE A.6　Length.

1 angstrom (Å) =	1 centimeter (cm) =	1 inch (in.) =	1 foot (ft) =
1 Å	10^8 Å	2.54×10^8 Å	3.048×10^9 Å
10^{-8} cm	1 cm	2.54 cm	30.48 cm
3.937×10^{-9} in.	0.3937 in.	1 in.	12 in.
3.281×10^{-10} ft	3.281×10^{-2} ft	8.333×10^{-2} ft	1 ft
10^{-10} m	0.01 m	0.0254 m	0.3048 m
10^{-13} km	10^{-5} km	2.54×10^{-5} km	3.048×10^{-4} km
6.214×10^{-14} mile	6.214×10^{-6} mile	1.578×10^{-5} mile	1.894×10^{-4} mile

1 meter (m) =	1 kilometer (km) =	1 mile =
10^{10} Å	10^{13} Å	1.609×10^{13} Å
100 cm	10^5 cm	1.609×10^5 cm
39.37 in.	3.937×10^4 in.	6.336×10^4 in.
3.281 ft	3281 ft	5280 ft
1 m	1000 m	1609 m
0.001 km	1 km	1.609 km
6.214×10^{-4} mile	0.6214 mile	1 mile

Other units of length:

1 fermi = 10^{-15} m, often used in referring to sizes of nuclei.

1 micron (μ) = 10^{-6} m

1 millimeter (mm) = 10^{-3} m

1 nautical mile = 1.852 km = 1.151 statute miles

1 light-year = 9.461×10^{15} m, the distance light travels in one year.

Note. The *mile* referred to in this table and everywhere in this book is the common "statute mile" (5280 ft).

TABLE A.7　Mass.

1 atomic mass unit (amu) =	1 gram (g) =	1 pound (lb)* =
1 amu	6.022×10^{23} amu	2.732×10^{26} amu
1.661×10^{-24} g	1 g	453.6 g
3.661×10^{-27} lb	2.205×10^{-3} lb	1 lb
1.661×10^{-27} kg	0.001 kg	0.4536 kg
1.83×10^{-30} ton	1.102×10^{-6} ton	5×10^{-4} ton
1.661×10^{-30} metric ton	10^{-6} metric ton	4.536×10^{-4} metric ton

1 kilogram (kg) =	1 ton*† =	1 metric ton (t or tonne)† =
6.022×10^{26} amu	5.463×10^{29} amu	6.022×10^{29} amu
1000 g	9.072×10^5 g	10^6 g
2.205 lb	2000 lb	2205 lb
1 kg	907.2 kg	1000 kg
1.102×10^{-3} ton	1 ton	1.102 tons
0.001 metric ton	0.9072 metric ton	1 metric ton

Other units of mass: 1 ounce = 1/16 lb = 2.835×10^{-2} kg; 1 slug = 32.17 lb.

*The pound and ton are used as units of both force and mass. A force of 1 lb is the force exerted by gravity on an object whose mass is 1 lb at a location where the acceleration due to gravity has its standard value. Other units of mass such as the kilogram are also occasionally used as units of force in this fashion.

†The *ton* in this table is the commonly used "short ton" (2000 lb), as opposed to the "long ton" (2240 lb). The *metric* ton (1000 kg), 10% greater than the short ton, is often used in discussions of energy resources.

TABLE A.8 Mass-energy equivalence.

NOTE: This table can be used to calculate the energy released in a process in which the mass decreases by a known amount. This table cannot be used directly to calculate, for example, the energy available from a given quantity of U^{235}, because in the fission of 1 kg of U^{235}, the decrease in mass is much less than 1 kg.

1 electron mass	1 atomic mass unit (amu)	1 gram (g)	1 kilogram (kg)
0.511 MeV	931.5 MeV	5.61×10^{26} MeV	5.61×10^{29} MeV
8.187×10^{-14} J	1.492×10^{-10} J	8.988×10^{13} J	8.988×10^{16} J
7.765×10^{-17} BTU	1.415×10^{-13} BTU	8.524×10^{10} BTU	8.524×10^{13} BTU
2.274×10^{-20} kWh	4.146×10^{-17} kWh	2.497×10^{7} kWh	2.497×10^{10} kWh
7.765×10^{-35} Q	1.415×10^{-31} Q	8.524×10^{-8} Q	8.524×10^{-5} Q

1 ton	1 MeV	1 joule (J)	1 BTU
5.089×10^{32} MeV	1.957 electron masses	1.221×10^{13} electron masses	1.288×10^{16} electron masses
8.153×10^{19} J	1.074×10^{-3} amu	6.701×10^{9} amu	7.065×10^{12} amu
7.733×10^{16} BTU	1.783×10^{-27} g	1.113×10^{-14} g	1.173×10^{-11} g
2.265×10^{13} kWh	1.783×10^{-30} kg	1.113×10^{-17} kg	1.173×10^{-14} kg
7.733×10^{-2} Q	1.965×10^{-33} ton	1.226×10^{-20} ton	1.293×10^{-17} ton

1 kilowatt-hour (kWh)	1 Q
4.397×10^{19} electron masses	1.288×10^{34} electron masses
2.412×10^{16} amu	7.065×10^{30} amu
4.006×10^{-8} g	1.173×10^{7} g
4.006×10^{-11} kg	1.173×10^{4} kg
4.415×10^{-14} ton	12.93 tons

TABLE A.9 Power.

1 BTU per day =	1 kilowatt-hour per year (kWh/yr) =	1 watt (W) =	1 kilowatt (kW) =
1 BTU/day	9.348 BTU/day	81.95 BTU/day	8.195×10^{4} BTU/day
0.107 kWh/yr	1 kWh/yr	8.766 kWh/yr	8766 kWh/yr
0.0122 W	0.1141 W	1 W	1000 W
1.22×10^{-5} kW	1.141×10^{-4} kW	0.001 kW	1 kW
1.22×10^{-8} MW	1.141×10^{-7} MW	10^{-6} MW	0.001 MW
1.22×10^{-11} GW	1.141×10^{-10} GW	10^{-9} GW	10^{-6} GW
3.652×10^{-13} mQ/yr	3.413×10^{-12} mQ/yr	2.993×10^{-11} mQ/yr	2.993×10^{-8} mQ/yr
3.652×10^{-16} Q/yr	3.413×10^{-15} Q/yr	2.993×10^{-14} Q/yr	2.993×10^{-11} Q/yr

1 megawatt (MW) =	1 gigawatt (GW) =	1 milli-Q per year (mQ/yr) =	1 Q per year (Q/yr) =
8.195×10^{7} BTU/day	8.195×10^{10} BTU/day	2.738×10^{12} BTU/day	2.738×10^{15} BTU/day
8.766×10^{6} kWh/yr	8.766×10^{9} kWh/yr	2.929×10^{11} kWh/yr	2.929×10^{14} kWh/yr
10^{6} W	10^{9} W	3.341×10^{10} W	3.341×10^{13} W
1000 kW	10^{6} kW	3.341×10^{7} kW	3.341×10^{10} kW
1 MW	1000 MW	3.341×10^{4} MW	3.341×10^{7} MW
0.001 GW	1 GW	33.41 GW	3.341×10^{4} GW
2.993×10^{-5} mQ/yr	2.993×10^{-2} mQ/yr	1 mQ/yr	1000 mQ/yr
2.993×10^{-8} Q/yr	2.993×10^{-5} Q/yr	0.001 Q/yr	1 Q/yr

Other units of power: 1 horsepower (hp) = 746 W.

TABLE A.10   Pressure.

1 newton per square meter (N/m²) =	1 pound per square foot (lb/ft²) =	1 pound per square inch (lb/in.² or psi) =	1 atmosphere (atm) =
1 N/m²	47.88 N/m²	6895 N/m²	1.013×10^5 N/m²
2.089×10^{-2} lb/ft²	1 lb/ft²	144 lb/ft²	2116 lb/ft²
1.45×10^{-4} lb/in.²	6.944×10^{-3} lb/in.²	1 lb/in.²	14.7 lb/in.²
9.869×10^{-6} atm	4.725×10^{-4} atm	6.805×10^{-2} atm	1 atm

Other units of pressure:

1 bar $= 10^5$ N/m² $= 0.9869$ atm.

Pressure is often measured by giving the height of a column of water or mercury that exerts such a pressure at its base: 1 atm = 76 cm of mercury = 33.9 ft of water.

TABLE A.11   Temperature.

Temperature intervals	1°C (Celsius or centigrade) = 1°K (Kelvin or absolute) = 1.8°F (Fahrenheit)
	$1°F = \frac{5}{9} °K = \frac{5}{9} °C.$
Correspondence between temperature scales	On the various scales, the normal freezing point of water is: 32°F = 0°C = 273.15°K.
	Absolute or Kelvin temperatures are often written without the ° symbol and read as "kelvins." Thus the normal freezing point of water is 273.15 K or 273.15 kelvins.
Conversions between temperature scales	From the relationships between the temperature intervals and the given values of the freezing point of water, it follows that:

$$T\ (°F) = 32 + 1.8 \times T(°C)$$
$$T\ (°C) = \tfrac{5}{9} \times [T(°F) - 32]$$
$$T\ (°K) = 273.15 + T(°C)$$

	°C	°F	°K
Absolute zero	−273.15	−459.67	0
Normal freezing point of water	0	32	273.15
Normal boiling point of water	100	212	373.15

TABLE A.12   Time.

1 second (sec or s) =	1 minute (min) =	1 hour (hr or h) =	1 day (d) =	1 year (yr or y) =
1 sec	60 sec	3600 sec	8.64×10^4 sec	3.156×10^7 sec
1.667×10^{-2} min	1 min	60 min	1440 min	5.259×10^5 min
2.778×10^{-4} hr	1.667×10^{-2} hr	1 hr	24 hr	8766 hr
1.157×10^{-5} day	6.944×10^{-4} day	4.167×10^{-2} day	1 day	365.2 days
3.169×10^{-8} yr	1.901×10^{-6} yr	1.141×10^{-4} yr	2.738×10^{-3} yr	1 yr

TABLE A.13   Velocity.

1 centimeter per second (cm/sec) =	1 kilometer per hour (km/hr) =	1 foot per second (ft/sec) =	1 mile per hour (mile/hr or mph) =
1 cm/sec	27.78 cm/sec	30.48 cm/sec	44.7 cm/sec
0.036 km/hr	1 km/hr	1.097 km/hr	1.609 km/hr
3.281×10^{-2} ft/sec	0.9113 ft/sec	1 ft/sec	1.467 ft/sec
2.237×10^{-2} mile/hr	0.6214 mile/hr	0.6818 mile/hr	1 mile/hr
0.01 m/sec	0.2778 m/sec	0.3048 m/sec	0.447 m/sec
10^{-5} km/sec	2.778×10^{-4} km/sec	3.048×10^{-4} km/sec	4.47×10^{-4} km/sec
6.214×10^{-6} mile/sec	1.726×10^{-4} mile/sec	1.894×10^{-4} mile/sec	2.778×10^{-4} mile/sec

1 meter per second (m/sec) =	1 kilometer per second (km/sec) =	1 mile per second (mile/sec) =
100 cm/sec	10^5 cm/sec	1.609×10^5 cm/sec
3.6 km/hr	3600 km/hr	5794 km/hr
3.281 ft/sec	3281 ft/sec	5280 ft/sec
2.237 miles/hr	2237 miles/hr	3600 miles/hr
1 m/sec	1000 m/sec	1609 m/sec
0.001 km/sec	1 km/sec	1.609 km/sec
6.214×10^{-4} mile/sec	0.6214 mile/sec	1 mile/sec

TABLE A.14   Volume.

1 cubic centimeter (cm³ or cc) =	1 cubic inch (in.³) =	1 liter (l) =
1 cm³	16.39 cm³	1000 cm³
6.102×10^{-2} in.³	1 in.³	61.02 in.³
0.001 liter	1.639×10^{-2} liter	1 liter
2.642×10^{-4} gal	4.329×10^{-3} gal	0.2642 gal
3.531×10^{-5} ft³	5.787×10^{-4} ft³	3.531×10^{-2} ft³
6.29×10^{-6} barrel	1.031×10^{-4} barrel	6.29×10^{-3} barrel
10^{-6} m³	1.639×10^{-5} m³	0.001 m³
10^{-15} km³	1.639×10^{-14} km³	10^{-12} km³
2.399×10^{-16} mile³	3.931×10^{-15} mile³	2.399×10^{-13} mile³

1 gallon (gal) =	1 cubic foot (ft³) =	1 barrel (bbl) =
3785 cm³	2.832×10^{4} cm³	1.59×10^{5} cm³
231 in.³	1728 in.³	9702 in.³
3.785 liters	28.32 liters	159 liters
1 gal	7.481 gal	42 gal
0.1337 ft³	1 ft³	5.615 ft³
2.381×10^{-2} barrel	0.1781 barrel	1 barrel
3.785×10^{-3} m³	2.832×10^{-2} m³	0.159 m³
3.785×10^{-12} km³	2.832×10^{-11} km³	1.59×10^{-10} km³
9.082×10^{-13} mile³	6.794×10^{-12} mile³	3.814×10^{-11} mile³

1 cubic meter (m³) =	1 cubic kilometer (km³) =	1 cubic mile (mile³) =
10^{6} cm³	10^{15} cm³	4.168×10^{15} cm³
6.102×10^{4} in.³	6.102×10^{13} in.³	2.544×10^{14} in.³
1000 liters	10^{12} liters	4.168×10^{12} liters
264.2 gal	2.642×10^{11} gal	1.101×10^{12} gal
35.31 ft³	3.531×10^{10} ft³	1.472×10^{11} ft³
6.29 barrels	6.29×10^{9} barrels	2.622×10^{10} barrels
1 m³	10^{9} m³	4.168×10^{9} m³
10^{-9} km³	1 km³	4.168 km³
2.399×10^{-10} mile³	0.2399 mile³	1 mile³

Other units of volume:

1 British gallon = 1.2 U.S. gal

1 cord (of wood) = 128 ft³

1 acre-foot = 1233 m³, often used in measuring quantities of water. (1 acre-foot is the volume of water that covers an area of 1 acre to a depth of 1 ft.)

1 quart = 1/4 gal = 0.9464 liter; 1 pint = 1/2 qt = 0.4732 liter. The pint, quart and gallon are U.S. units of *liquid* measurement; different "dry" measures are occasionally used.

The barrel used here is the 42-gallon barrel used in measuring petroleum; many other sizes of "barrels" are in use for other special purposes.

B
Abbreviations and Symbols, Decimal Multiples, and Geometrical Formulas

TABLE B.1   Abbreviations and symbols.

A	mass number or nucleon number	EPE	electrical potential energy	kV	kilovolt	R	resistance
A	ampere	eV	electron-volt	kW	kilowatt	rad	radiation absorbed dose
Å	angstrom unit	F	force	kWh	kilowatt-hour		
AC	alternating current	°F	degree-Fahrenheit	l	liter	rem	roentgen equivalent man
A h	ampere-hour	ft	foot	lb	pound		
amu	atomic mass unit	g	gram	LMFBR	liquid-metal fast breeder reactor	rms	root-mean-square
atm	atmosphere	g	acceleration due to gravity	m	meter	S_o	solar constant
B	magnetic field strength			m	mass	sec	second
		G	universal gravitational constant	M	mega	t	triton
bbl	barrel			MBTU	millions of BTU's	T	tritium
BTU	British thermal unit	G	giga	MeV	million electron-volts	T	temperature
BWR	boiling water reactor	gal	gallon	min	minute	TE	thermal energy
		GNP	gross national product	MKS	meter-kilogram-second	v	velocity or speed
c	speed of light					V	volt
C	heat capacity	GPE	gravitational potential energy	mm	millimeter	V	voltage
C	coulomb			mph	miles per hour	W	watt
°C	degree-Celsius or degree-centigrade	GW	gigawatt	mQ	milli-Q	W	work
		H	heat	mrem	millirem	yr	year
cal	calorie	h	Planck's constant	MW	megawatt	Z	atomic number or proton number
Cal	kilocalorie	hr	hour	MW(e)	megawatt-electrical		
CE	chemical energy	Hz	hertz (cycles per second)	MW(t)	megawatt-thermal		
cm	centimeter			n	neutron	α	alpha-particle or earth's albedo
CP	coefficient of performance	I	current	N	neutron number		
		in.	inch	N	newton	β	beta-particle
d	deuteron	J	joule	N_A	Avogadro's number	β^-	electron
D	deuterium	k	kilo	NGL	natural gas liquids	β^+	positron
DC	direct current	k	Boltzmann's constant or thermal conductance	oz	ounce	γ	gamma-ray
DD	degree-day			p	proton	Δ	calculate the *change* in the following quantity
e	electron or electronic charge	k_e	Coulomb's law constant	P	power		
				psi	pounds per square inch		
e^-	electron	K	K-capture	PWR	pressurized water reactor	ϵ	efficiency
e^+	positron	°K	degree-Kelvin			λ	wavelength
E.E.R.	energy efficiency ratio	kcal	kilocalorie	q	charge	ν	frequency
		KE	kinetic energy	Q	an energy unit, 10^{18} BTU	ρ	resistivity
E	energy	kg	kilogram			σ	Stefan-Boltzmann constant
ε	electric field strength	km	kilometer	R	gas constant	τ	containment time
emf	electromotive force						

TABLE B.2 Prefixes and abbreviations for decimal multiples and sub-multiples.

Multiple	Prefix	Abbreviation	Example
10^{-12}	pico	p	1 picosecond = 10^{-12} sec
10^{-9}	nano	n	1 nanovolt = 10^{-9} V
10^{-6}	micro	μ	1 μsec = 10^{-6} sec
10^{-3}	milli	m	1 mm = 1 millimeter = 10^{-3} m
10^{-2}	centi	c	1 cm = 1 centimeter = 0.01 m
10^{3}	kilo	k	1 kg = 1 kilogram = 1000 g
10^{6}	mega	M	1 MW = 1 megawatt = 10^{6} W
10^{9}	giga	G	1 GW = 1 gigawatt = 10^{9} W*
10^{12}	tera	T	1 TBTU = 1 teraBTU = 10^{12} BTU*

*In common American usage, 10^{9} W = 1 *billion* watts and 10^{12} BTU = 1 *trillion* BTU. The terms thousand and million for 10^{3} and 10^{6} are universally understood. In the United States, 1 billion = 10^{9} (1000 million), 1 trillion = 10^{12} (1000 billion), etc., and whenever these terms are used in this book, it is with these meanings. In some countries, however, 1 billion means 10^{12} (1 million million), 1 trillion = 10^{18}, etc., so that caution should be exercised in comparing American and foreign sources in which these terms are used.

TABLE B.3 Geometrical formulas.

	Values	
Circle of radius r	Circumference = $2\pi r$	Area = πr^2
Sphere of radius r	Surface area = $4\pi r^2$	Volume = $\frac{4}{3}\pi r^3$
Numerical value of π	$\pi \simeq 3.1416$	

C
Physical and Chemical Data

TABLE C.1   Physical constants.

	Values
Universal gravitational constant	$G = 6.673 \times 10^{-11}$ m³/kg-sec²
Acceleration due to gravity	$g = 9.80665$ m/sec² ("standard" value)
	$g = 9.8$ m/sec² $= 980$ cm/sec² $= 32$ ft/sec² (for most purposes)
Avogadro's number	$N_A = 6.022 \times 10^{23}$ molecules/mole
Gas constant	$R = kN_A = 8.314$ J/mole-°K
Boltzmann's constant	$k = R/N_A = 1.381 \times 10^{-23}$ J/°K
Black-body radiation	
Stefan-Boltzmann constant (The power radiated by an ideal black surface of area A (in square meters) at an absolute temperature T, is given by $P = \sigma AT^4$.)	$\sigma = 5.670 \times 10^{-8}$ W/m²-°K⁴
Peak wavelength (The equation gives the relationship between the wavelength at the peak of the spectral distribution of black-body radiation, in meters, and the temperature, in °K.)	$\lambda_{max} = 0.0029/T$
Planck's constant	$h = 6.626 \times 10^{-34}$ J-sec
Electronic charge (elementary unit of charge)	$e = 1.602 \times 10^{-19}$ coulombs (C)
Coulomb's law constant (The electrical force between two point charges separated by distance r, with q_1 and q_2 in coulombs, r in meters, and F in newtons, is: $F = k_e q_1 q_2 / r^2$.)	$k_e = 9 \times 10^9$ N-m²/C² (approximate value)
	$k_e = 8.988 \times 10^9$ N-m²/C² (more precise value)
Speed of light in vacuum (The speed of light in air is the same to within 0.1%.)	$c = 3 \times 10^8$ m/sec $= 3 \times 10^{10}$ cm/sec $= 186{,}000$ miles/sec (approximate values)
	$c = 2.998 \times 10^8$ m/sec (more precise value)
Speed of sound in dry air at standard temperature and pressure ($T = 0°C$, $p = 1$ atm)	331 m/sec $= 1087$ ft/sec $= 741$ miles/hr
Density of dry air at standard temperature and pressure	1.29 kg/m³ $= 1.29 \times 10^{-3}$ g/cm³
Atomic mass unit	1 amu $= 1.661 \times 10^{-27}$ kg $= 1.661 \times 10^{-24}$ g

TABLE C.1 (continued)

Masses for particles:

Particle	Mass		
	(amu)	(kg)	(g)
electron	5.48593×10^{-4}	9.110×10^{-31}	9.110×10^{-28}
proton	1.0072766	1.6726×10^{-27}	1.6726×10^{-24}
neutron	1.0086652	1.6749×10^{-27}	1.6749×10^{-24}
hydrogen atom (H¹)	1.0078252	1.6735×10^{-27}	1.6735×10^{-24}

Sizes of atoms and nuclei

Atomic radii	Typically in the range of 1–2 Å for all atoms. (The "radius" of an atom is not a quantity that can be given a precise definition.)
Nuclear radii	Approximate values of nuclear radii are given by the expression:

$$r \simeq r_o A^{\frac{1}{3}},$$

with $r_o = 1.4 \times 10^{-15}$ m and A the atomic mass number (or nucleon number). Thus from the lightest nuclei ($A = 1$) to the heaviest ($A \simeq 240$), r varies from about 1.4×10^{-15} m to 8.7×10^{-15} m.

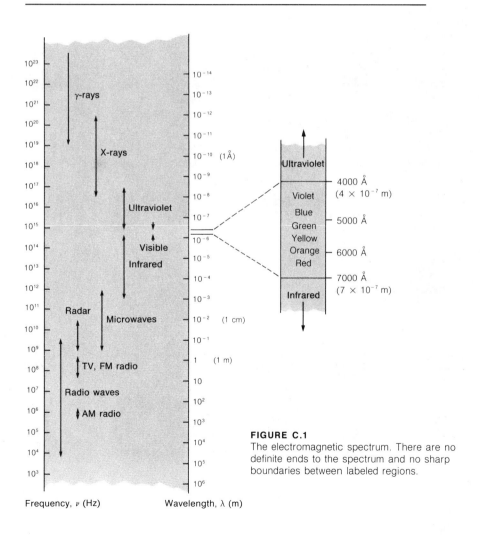

FIGURE C.1
The electromagnetic spectrum. There are no definite ends to the spectrum and no sharp boundaries between labeled regions.

TABLE C.2 Periodic table of the elements.

For each element, the number in the upper left-hand corner is Z (the atomic number or proton number). The number in the upper right-hand corner is the average mass in atomic mass units of the atoms in a sample containing the various isotopes in their usual natural abundance; for those elements not found in nature or for those such as polonium, which are only found as decay products of long-lived elements, the mass number or nucleon number (A) of the isotope with the longest known half-life is given in parentheses.

1 1.008 **H** Hydrogen							
3 6.939 **Li** Lithium	4 9.012 **Be** Beryllium						
11 22.99 **Na** Sodium	12 24.31 **Mg** Magnesium						
19 39.102 **K** Potassium	20 40.08 **Ca** Calcium	21 44.96 **Sc** Scandium	22 47.90 **Ti** Titanium	23 50.94 **V** Vanadium	24 52.00 **Cr** Chromium	25 54.94 **Mn** Manganese	26 55.85 **Fe** Iron
37 85.47 **Rb** Rubidium	38 87.62 **Sr** Strontium	39 88.91 **Y** Yttrium	40 91.22 **Zr** Zirconium	41 92.91 **Nb** Niobium	42 95.94 **Mo** Molybdenum	43 (97) **Tc** Technetium	44 101.1 **Ru** Ruthenium
55 132.91 **Cs** Cesium	56 137.34 **Ba** Barium	57–71 Lanthanide series*	72 178.5 **Hf** Hafnium	73 180.95 **Ta** Tantalum	74 183.85 **W** Tungsten	75 186.2 **Re** Rhenium	76 190.2 **Os** Osmium
87 (223) **Fr** Francium	88 (226) **Ra** Radium	89–103 Actinide series†	104 (257)	105 (260)	106		

(continued columns)

27 58.93 **Co** Cobalt
45 102.91 **Rh** Rhodium
77 192.2 **Ir** Iridium

*Lanthanide Series

57 138.9 **La** Lanthanum	58 140.1 **Ce** Cerium	59 140.9 **Pr** Praseodymium	60 144.2 **Nd** Neodymium	61 (145) **Pm** Promethium	62 150.4 **Sm** Samarium	63 152.0 **Eu** Europium

†Actinide Series

89 (227) **Ac** Actinium	90 232.0 **Th** Thorium	91 (231) **Pa** Protactinium	92 238.0 **U** Uranium	93 (237) **Np** Neptunium	94 (244) **Pu** Plutonium	95 (243) **Am** Americium

2 4.003 **He** Helium	

5 10.81 **B** Boron	6 12.01 **C** Carbon	7 14.01 **N** Nitrogen	8 15.999 **O** Oxygen	9 19.00 **F** Fluorine	10 20.18 **Ne** Neon
13 26.98 **Al** Aluminum	14 28.09 **Si** Silicon	15 30.97 **P** Phosphorus	16 32.06 **S** Sulfur	17 35.45 **Cl** Chlorine	18 39.95 **Ar** Argon

28 58.71 **Ni** Nickel	29 63.54 **Cu** Copper	30 65.37 **Zn** Zinc	31 69.72 **Ga** Gallium	32 72.59 **Ge** Germanium	33 74.92 **As** Arsenic	34 78.96 **Se** Selenium	35 79.91 **Br** Bromine	36 83.80 **Kr** Krypton
46 106.4 **Pd** Palladium	47 107.87 **Ag** Silver	48 112.4 **Cd** Cadmium	49 114.8 **In** Indium	50 118.7 **Sn** Tin	51 121.8 **Sb** Antimony	52 127.6 **Te** Tellurium	53 126.9 **I** Iodine	54 131.3 **Xe** Xenon
78 195.1 **Pt** Platinum	79 196.97 **Au** Gold	80 200.6 **Hg** Mercury	81 204.4 **Tl** Thallium	82 207.2 **Pb** Lead	83 209.0 **Bi** Bismuth	84 (209) **Po** Polonium	85 (210) **At** Astatine	86 (222) **Rn** Radon

64 157.3 **Gd** Gadolinium	65 158.9 **Tb** Terbium	66 162.5 **Dy** Dysprosium	67 164.9 **Ho** Holmium	68 167.3 **Er** Erbium	69 168.9 **Tm** Thulium	70 173.0 **Yb** Ytterbium	71 175.0 **Lu** Lutetium
96 (247) **Cm** Curium	97 (247) **Bk** Berkelium	98 (251) **Cf** Californium	99 (254) **Es** Einsteinium	100 (257) **Fm** Fermium	101 (258) **Md** Mendelevium	102 (255) **No** Nobelium	103 (256) **Lw** Lawrencium

TABLE C.3. Masses of some selected atoms and particles.

Except for the electron, proton, and neutron, the listed masses (in atomic mass units) are those of complete neutral atoms (the nucleus together with the appropriate number of electrons). Many more nuclear species are known than those listed here. For the elements that occur naturally, the natural abundance of each listed isotope is given, the percentage of atoms in a typical sample that have the particular mass number. (Abundances are not given for elements such as polonium that occur naturally but only as the decay products of nuclear species with longer half-lives.) For unstable nuclei, the half-life (in seconds, minutes, hours, days, or years) is given, together with the modes of decay; α, β^-, β^+ and K denote α-decay, ordinary β-decay, positron emission, and K-capture. In most decays, γ-rays are also emitted.

Element or particle	Symbol	Z (atomic number or proton number)	A (mass number or nucleon number)	Mass (atomic mass units)	Percentage abundance	Half-life	Common modes of decay
Electron	e, β^-			5.48593×10^{-4}			
Proton	p			1.0072766			
Neutron	n			1.0086652		11 min	β^-
Hydrogen	H	1	1	1.0078252	99.985		
			2	2.0141022	0.015		
			3	3.0160497		12.3 yr	β^-
Helium	He	2	3	3.0160297	0.00013		
			4	4.002603	99.9999		
Lithium	Li	3	6	6.015123	7.4		
			7	7.016004	92.6		
Boron	B	5	11	11.009305	80.2		
Carbon	C	6	12	12.000	98.9		
			13	13.003354	1.1		
			14	14.003242		5730 yr	β^-
Nitrogen	N	7	13	13.005738		10 min	β^+
			14	14.0030744	99.63		
			15	15.000108	0.37		
Oxygen	O	8	15	15.00307		124 sec	β^+
			16	15.99492	99.759		
			17	16.99913	0.037		
			18	17.99916	0.204		
Sodium	Na	11	23	22.989771	100		
Argon	Ar	18	40	39.962384	99.6		
Potassium	K	19	40	39.964	0.01	1.3×10^9 yr	K, β^-, β^+
			41	40.961827	6.88		
Calcium	Ca	20	40	39.962592	97		
			41	40.962279		7.7×10^4 yr	K
Scandium	Sc	21	41	40.96925		0.6 sec	β^+
Iron	Fe	26	56	55.934934	91.7		
Nickel	Ni	28	58	57.935336	67.9		
			64	63.92796	1.1		
Copper	Cu	29	58	57.94454		3.2 sec	β^+
			63	62.92959	69.1		
			64	63.929757		12.8 hr	K, β^-, β^+
			65	64.92779	30.9		

TABLE C.3 (continued)

Element or particle	Symbol	Z (atomic number or proton number)	A (mass number or nucleon number)	Mass (atomic mass units)	Percentage abundance	Half-life	Common modes of decay
Zinc	Zn	30	64	63.92914	48.9		
			65	64.92923		244 days	K, β^+
Gallium	Ga	31	65	64.93273		15 min	K, β^+
Germanium	Ge	32	65	64.94		1.5 min	β^+
Krypton	Kr	36	85	84.912537		10.8 yr	β^-
Rubidium	Rb	37	85	84.9118	72.2		
			90	89.9148		2.6 min	β^-
Strontium	Sr	38	90	89.90775		28.1 yr	β^-
Yttrium	Y	39	90	89.90716		64.2 hr	β^-
			97	96.918		1.1 sec	β^-
Zirconium	Zr	40	90	89.9047	51.5		
			97	96.91097		17 hr	β^-
Niobium	Nb	41	97	96.9081		72 min	β^-
Molybdenum	Mo	42	97	96.906023	9.5		
Silver	Ag	47	107	106.90509	51.8		
			108	107.90595		2.4 min	β^-
Cadmium	Cd	48	108	107.90419	0.9		
Neodymium	Nd	60	142	141.90777	27.1		
Samarium	Sm	62	146	145.9131		10^8 yr	α
Gold	Au	79	197	196.96655	100		
Thallium	Tl	81	207	206.97745		4.8 min	β^-
Lead	Pb	82	206	205.97447	23.6		
Bismuth	Bi	83	210	209.98413		5 days	β^-, α
			211	210.9873		2.1 min	β^-, α
Polonium	Po	84	210	209.98288		138 days	α
Radium	Ra	88	226	226.02544		1600 yr	α
Thorium	Th	90	232	232.03808	100	1.4×10^{10} yr	α
			233	233.0416		22.2 min	β^-
Protactinium	Pa	91	233	233.04027		27 days	β^-
Uranium	U	92	233	233.03965		1.6×10^5 yr	α
			235	235.04394	0.72	7×10^8 yr	α
			238	238.05082	99.27	4.5×10^9 yr	α
			239	239.05433		23.5 min	β^-
Neptunium	Np	93	239	239.05295		2.35 days	β^-
Plutonium	Pu	94	239	239.05218		2.44×10^4 yr	α

TABLE C.4. Values of various temperatures on the
Fahrenheit, Celsius and Kelvin scales.

	°F	°C	°K
Absolute zero	−460	−273	0
Liquid helium	−452	−269	4.2
Liquid hydrogen	−423	−253	20
Liquid nitrogen	−320	−196	77
Liquid oxygen	−297	−183	90
Lowest recorded weather temperature	−125	−87	186
"Dry ice" (solid CO_2)	−110	−79	194
Mercury freezes	−38	−39	234
Water freezes	32	0	273
Room temperature	70	21	294
Body temperature	98.6	37	310
Highest recorded weather temperature	136	58	331
Water boils	212	100	373
Lead melts	621	327	600
Steam temperature in a nuclear power plant	660	350	620
Steam temperature in a fossil-fuel power plant	930	500	770
Uranium fuel rod (interior temperature)	4000	2200	2500
Light bulb filament	4600	2500	2800
Tungsten melts	6170	3410	3683
Sun—surface	10,000	5500	5800
Sun—interior	2.7×10^7	1.5×10^7	1.5×10^7
Deuterium-Tritium fusion (ignition temperature)	7×10^7	4×10^7	4×10^7
Deuterium-Deuterium fusion (ignition temperature)	7×10^8	4×10^8	4×10^8

TABLE C.5. Electrical
resistivities (in ohm-meters)
of various materials at a
temperature of 20°C.

Material	Value, ρ (ohm-meters)
Silver	1.6×10^{-8}
Copper	1.7×10^{-8}
Gold	2.2×10^{-8}
Aluminum	2.7×10^{-8}
Tungsten	5.6×10^{-8}
Brass	7×10^{-8}
Iron	9.8×10^{-8}
Lead	21×10^{-8}
Uranium	26×10^{-8}
Mercury	96×10^{-8}
Nichrome*	100×10^{-8}
Sea water†	0.2
Distilled water	10^4
Bakelite	10^{11}
Hard rubber	10^{11}
Nylon	10^{13}
Mica	10^{14}
Glass	10^{14}

*Nichrome is an alloy often used in heating
elements of toasters and other appliances.

†The resistivities of sea water, bakelite, rubber,
nylon, mica and glass vary considerably from one
sample to another; listed values are typical.

TABLE C.6. Specific heat capacities (at constant
pressure) of various substances, near room temperature.

	cal/g-°C	J/kg-°C
Water	1.0	4184
Ice	0.5	2100
Aluminum	0.22	903
Iron	0.11	449
Copper	0.092	385
Lead	0.031	129
Gold	0.031	129
Brass	0.09	375
Brick	0.2	800
Cement	0.15	600
Rocks	0.2	800
Dirt (dry)	0.2	800
Wood	0.3-0.6	1200-2500
Glass	0.1-0.2	400-800
Bakelite	0.35	1400
Paper	0.3	1300
Gasoline	0.5	2100
Oxygen	0.22	918
Nitrogen	0.25	1040
Air (dry)	0.24	1000
Hydrogen	3.42	14175
Helium	1.24	5200

TABLE C.7. Latent heats of water.

	cal/g	J/kg
Latent heat of fusion (ice to water)	79.71	3.335×10^5
Latent heat of vaporization (water to steam)	539.6	2.257×10^6

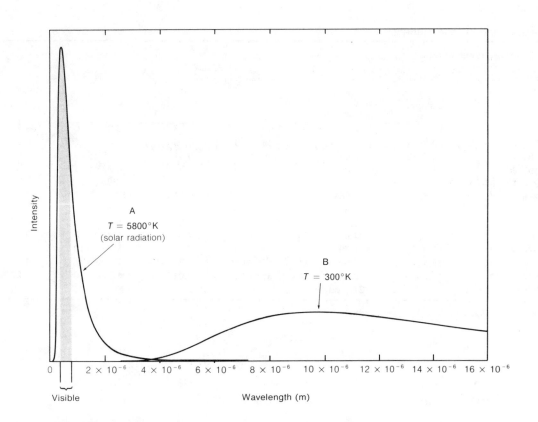

FIGURE C.2. Distribution of the blackbody radiation from an object at a temperature of 5800 K (approximately equal to the surface temperature of the sun, curve A) and 300 K (approximately the temperature of the earth, curve B).

D
The Solar System

TABLE D.1 The sun, moon, and planets.

	Average radius of orbit		Orbital period (yr)	Mass		Radius	
	km	Relative to earth-sun distance		kg	Relative to earth's mass	km	Relative to earth's radius
Sun	—	—	—	1.99×10^{30}	3.33×10^5	6.96×10^5	109.2
Mercury	5.79×10^7	0.3871	0.2409	3.35×10^{23}	0.056	2.485×10^3	0.39
Venus	1.081×10^8	0.7233	0.6152	4.89×10^{24}	0.817	6.18×10^3	0.97
Earth	1.495×10^8 (92.9×10^6 miles)	1.0	1.0	5.98×10^{24}	1.0	6.371×10^3 (3959 miles)	1.0
Mars	2.278×10^8	1.524	1.881	6.46×10^{23}	0.108	3.377×10^3	0.53
Jupiter	7.778×10^8	5.203	11.862	1.90×10^{27}	318	7.13×10^4	11.19
Saturn	1.426×10^9	9.539	29.458	5.69×10^{26}	95.2	6.03×10^4	9.47
Uranus	2.868×10^9	19.182	84.013	8.73×10^{25}	14.6	2.35×10^4	3.69
Neptune	4.494×10^9	30.058	164.79	1.03×10^{26}	17.3	2.23×10^4	3.50
Pluto	5.896×10^9	39.439	247.69	5.4×10^{24}	0.9	7×10^3	1.1
Moon*	3.844×10^5 (2.389×10^5 miles)	0.00257	0.0747 (27.3 days)	7.35×10^{22}	0.012	1.72×10^3	0.27

*Data for the moon's orbit refer to its orbit around the earth

24

E
The Earth

TABLE E.1   The earth.

	Values
Average radius:	6371 km = 3959 miles
Mass:	5.98×10^{24} kg = 6.59×10^{21} tons
Volume:	1.083×10^{27} cm³ = 1.083×10^{12} km³
	= 2.6×10^{11} cubic miles
Average density:	5.52 g/cm³ = 345 lb/ft³
Surface area:	5.1×10^8 km² = 1.97×10^8 square miles
Land (29%):	1.49×10^8 km² = 5.75×10^7 square miles
Oceans (71%):	3.61×10^8 km² = 1.39×10^8 square miles
Average distance to sun:	1.495×10^8 km = 92.9×10^6 miles
(The actual distance varies by about 1.7% above and below this value during the year. The earth is closest to the sun on about January 1 and farthest on about July 1.)	
Orbital velocity of earth around sun:	29.8 km/sec = 18.5 miles/sec
Velocity of a point on the equator due to rotation of earth about its axis:	0.465 km/sec = 0.289 miles/sec
Acceleration due to gravity, g	g = 9.8 m/sec² = 980 cm/sec²
(The "standard" value of g is 9.80665 m/sec². The value of g varies with elevation and location; the values listed, correct to two significant figures, are suitable for most purposes.)	= 32 ft/sec²
Average albedo (fraction of incident sunlight reflected)	0.34

TABLE E.2   The earth's continental crusts.

Depth: approximately 35 km
Average density: 2.7 g/cm³ = 170 lb/ft³

Element*	Average abundance† (in parts per million by weight)
Oxygen	464,000
Silicon	282,000
Aluminum	82,000
Iron	56,000
Calcium	41,000
Sodium	24,000
Magnesium	23,000
Potassium	21,000
Titanium	5,700
Hydrogen	1,400
Lithium	20
Thorium	9.6
Uranium	2.7

*The elements listed are the ten most abundant plus three others (lithium, thorium, and uranium), which are less common but which may be of special interest as energy sources.

†In practice, a particular element can be mined only from deposits of higher than average concentration.

TABLE E.3   The earth's atmosphere.

	Values
Normal atmospheric pressure at sea-level:	14.7 lb/in.^{2} = 1.013 × 10^5 N/m^2
Average height (altitude at which pressure is half of its value at sea-level):	5.5 km = 3.4 miles
Volume contained between earth's surface and altitude of 5.5 km:	2.8 × 10^9 km^3 = 6.7 × 10^8 cubic miles
Total mass:	5.14 × 10^{18} kg = 5.66 × 10^{15} tons
Total water content:	1.3 × 10^{16} kg = 1.43 × 10^{13} tons (equivalent to 1.3 × 10^4 km^3 of liquid), 0.001% of earth's total amount of water
Average density of dry air at sea level at standard temperature and pressure (T = 0°C, p = 1 atmosphere):	1.29 × 10^{-3} g/cm^3 = 1.29 kg/m^3
Velocity of sound in dry air at standard temperature and pressure:	331 m/sec = 1087 ft/sec = 741 miles/hr

Composition of the earth's atmosphere* (normal composition of clean dry air near sea level)

Substance	Percentage by number of molecules	Percentage by weight
Nitrogen (N$_2$)	78.084	75.52
Oxygen (O$_2$)	20.948	23.14
Argon (Ar)	0.934	1.29
Carbon dioxide (CO$_2$)	0.0314	0.0477
Neon (Ne)	0.0018	0.0013
Helium (He)	0.00052	0.00007
Methane (CH$_4$)	0.0002	0.0001
Krypton (Kr)	0.00011	0.0003
Hydrogen (H$_2$)	0.00005	0.000003
Nitrous oxide (N$_2$O)	0.00005	0.00008
Xenon (Xe)	0.000009	0.00004

*The atmosphere also contains water vapor as an important constituent with variable concentration. Concentrations of carbon dioxide and methane vary significantly from time to time and place to place. Small amounts of sulfur dioxide (SO$_2$), nitrogen dioxide (NO$_2$), ammonia (NH$_3$), carbon monoxide (CO), and iodine (I$_2$) are also found, and (especially in the upper atmosphere) so is ozone (O$_3$).

TABLE E.4   The earth's oceans.

	Values
Area (71% of the earth's surface area)	3.61×10^8 km^2 = 1.39×10^8 square miles
Volume (97.2% of all the earth's water)	1.32×10^9 km^3 = 3.17×10^8 cubic miles
Mass	1.36×10^{21} kg = 1.5×10^{18} tons
Average depth	3.7 km = 2.3 miles
Composition* (concentrations of the elements are given in g/m^3)†	
Oxygen	857,000
Hydrogen	108,000
Chlorine	19,000
Sodium	10,500
Magnesium	1,350
Sulfur	885
Calcium	400
Potassium	380
Bromine	65
Carbon	28
Lithium	0.17
Uranium	0.003
Thorium	less than 0.0000005

*Listed, in order of abundance are the ten most abundant elements, including hydrogen and oxygen, the elements that combine to form water (H_2O), as well as lithium, uranium and thorium, which are of interest as possible sources of energy. At least trace amounts of nearly all elements have been found in sea water.

†Since the mass of 1 m^3 of water is very nearly 10^6 g, these figures can also be interpreted as the concentrations in parts per million by weight.

TABLE E.5   The earth's supply of water.

Location	Water volume (km^3)	Percentage of total water	Equivalent depth*
Oceans	1.32×10^9	97.2	2.6 km
Icecaps and glaciers	2.9×10^7	2.14	57 m
Ground water and soil moisture	8.4×10^6	0.62	16.5 m
Fresh-water lakes	1.2×10^5	0.009	24 cm
Saline lakes and inland seas	10^5	0.007	20 cm
Atmosphere	1.3×10^4	0.001	2.5 cm
Rivers (amount flowing in rivers at any moment)	1.2×10^3	0.0001	0.24 cm
Total	1.36×10^9	100.0	2.7 km

*The depth if this amount of water were spread uniformly over the whole surface of the earth.

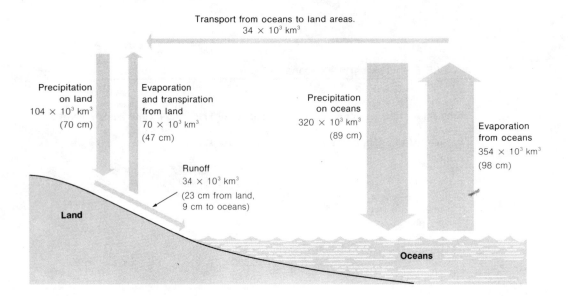

Transport from oceans to land areas.
34 × 10³ km³

Precipitation
on land
104 × 10³ km³
(70 cm)

Evaporation
and transpiration
from land
70 × 10³ km³
(47 cm)

Precipitation
on oceans
320 × 10³ km³
(89 cm)

Evaporation
from oceans
354 × 10³ km³
(98 cm)

Runoff
34 × 10³ km³
(23 cm from land,
9 cm to oceans)

Land

Oceans

FIGURE E.1
The earth's water cycle. Annual amounts of
water transported by precipitation,
evaporation, transpiration of plants and
animals, and river run-off. Values in
parentheses are the depths of the layers of
water that would result if the various amounts
of water were spread uniformly over the
total land area (1.49 × 10⁸ km²) or the total
ocean area (3.61 × 10⁸ km²). For example,
average annual precipitation is 70 cm on
land areas, and 89 cm on the oceans.

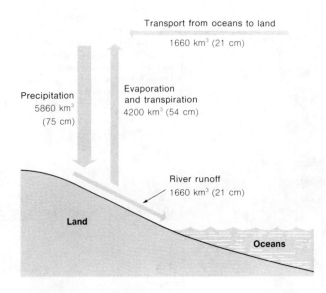

Transport from oceans to land
1660 km³ (21 cm)

Precipitation
5860 km³
(75 cm)

Evaporation
and transpiration
4200 km³ (54 cm)

River runoff
1660 km³ (21 cm)

Land

Oceans

FIGURE E.2
Water flow in the United States (48 contiguous states). Annual
amounts of water transported are given in cubic kilometers.
Numbers in parentheses are the depths of the layers of water
that would result if the various amounts of water were spread
uniformly over the total area of the 48 states.

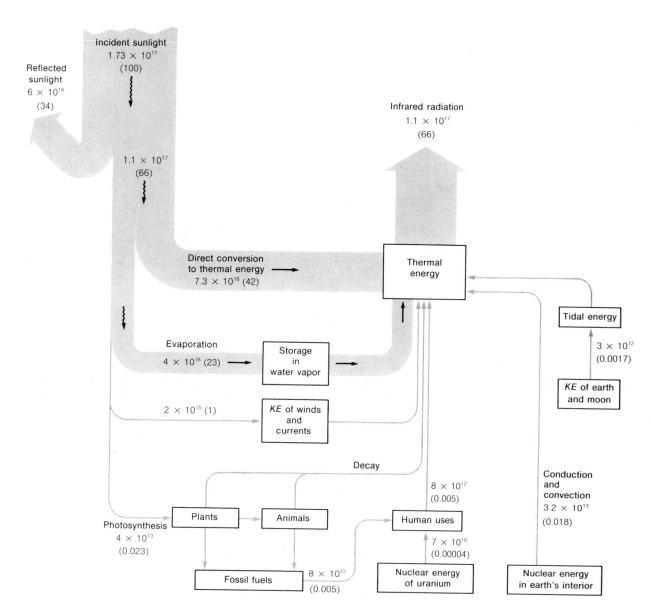

FIGURE E.3
The earth's energy flows, with the details of the exchange of radiation between the surface and the atmosphere omitted. Estimated values are given both in watts and (in parentheses) as percentages of the total power incident from the sun at the top of the atmosphere (1.73×10^{17} W).

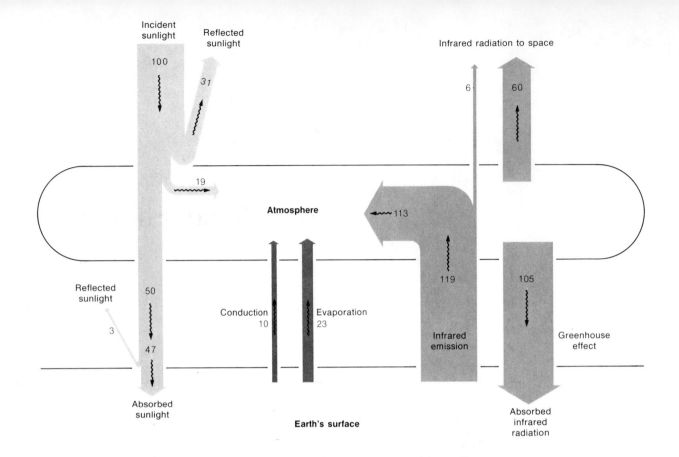

FIGURE E.4. Energy transfer between the sun, the atmosphere, the earth's surface, and the surrounding space. Numerical values are percentages of the solar power incident at the top of the atmosphere $(1.73 \times 10^{17}$ W). Light gray indicates short-wavelength solar radiation, medium gray shows long-wavelength infrared radiation, and dark gray shows nonradiative energy transfer.

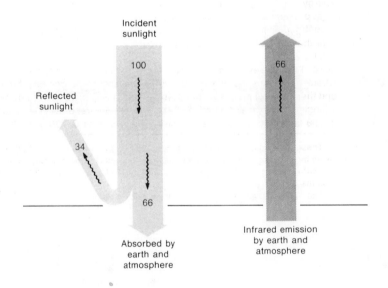

FIGURE E.5. Energy flows to and from the system composed of the earth together with its atmosphere (a simplified version of Figure E.4.)

F
Solar Energy

TABLE F.1   Solar energy data.

	Values
Total power radiated from the sun in all directions	3.8×10^{26} W
Power incident on the earth at the top of the atmosphere*	1.73×10^{17} W $= 5.45 \times 10^{24}$ J/yr $= 5174$ Q/yr
Solar constant (S_o)—power per unit area at the top of the earth's atmosphere, for a surface directly facing the sun*	$S_o = 1.353$ kW/m² (approximate rounded value: 1.4 kW/m²)
Amount incident at ground level per unit area, for a surface directly facing the sun (this amount varies with weather conditions and with the amount of atmosphere in the path; the value given here is typical for a time near noon on a clear and cloudless day)	1 kW/m²
Energy delivered to a horizontal surface (approximate average rate for the 48 contiguous states, averaged over all hours of the day and night and averaged over a full year)†	200 W/m²

NOTE: The langley. Solar energy data are frequently reported in *langleys* per day, where 1 langley is defined as 1 cal/cm². The langley is a unit of energy per unit area, and therefore the number of langleys per unit time is the average power per unit area: 1 langley/day = 0.484 W/m². El Paso, for instance, receives during June an average of 730 langleys/day, an average power per unit area of 730 × 0.484 = 353 W/m².

*These are average values for a whole year. During the year, the rate at which energy is received varies by 3.4% above and below these values, the highest value occurring near January 1 when the earth is closest to the sun, and the lowest value on about July 1 when the earth is farthest from the sun.

†See Table F.2 and Figure F.1 for more detailed data.

TABLE F.2 Average solar radiation for selected cities (incident radiation on a horizontal surface, averaged over all hours of day and night, W/m²).

	July	Aug.	Sept.	Oct.	Nov.	Dec.	Jan.	Feb.	Mar.	Apr.	May	June	Annual average
Atlanta, Ga.	257	246	201	166	130	102	106	140	184	236	258	272	192
Barrow, Alaska	208	123	56	20	0	0	0	18	87	184	248	256	100
Bismarck, N. Dak.	296	251	185	132	78	60	76	121	170	217	267	284	178
Boise, Idaho	324	275	221	152	88	60	69	113	164	235	284	309	191
Boston, Mass.	240	206	165	115	70	58	67	96	142	176	228	242	150
Caribou, Maine	246	218	161	102	53	51	66	111	178	194	229	232	153
Cleveland, Ohio	267	239	182	127	68	56	60	87	151	182	253	271	162
Columbia, Mo.	278	255	217	157	107	82	87	121	166	210	257	276	184
Dodge City, Kans.	311	287	239	184	138	113	123	153	202	256	275	315	216
El Paso, Tex.	324	309	278	224	178	151	160	209	266	317	346	353	260
Fresno, Calif.	323	293	243	182	117	77	90	143	212	264	308	337	216
Greensboro, N.C.	263	235	197	156	118	95	97	134	171	227	257	273	185
Honolulu, Hawaii	305	293	271	245	208	176	175	200	234	262	300	297	247
Indianapolis, Ind.	262	237	196	142	86	64	70	103	153	192	236	263	167
Little Rock, Ark.	270	250	214	167	118	91	96	127	173	220	256	272	188
Miami, Fla.	260	246	216	188	171	154	166	201	238	263	267	257	219
New York, N.Y.	251	238	175	127	77	62	71	102	151	183	220	255	159
Oklahoma City, Okla.	295	285	234	183	137	115	123	153	197	241	261	302	210
Omaha, Nebr.	275	252	192	142	96	80	99	134	172	224	248	272	182
Rapid City, S. Dak.	288	262	208	152	99	76	90	135	193	235	259	287	190
St. Cloud, Minn.	269	238	174	117	71	60	82	121	177	205	242	262	168
Salt Lake City, Utah	308	274	220	162	103	75	85	127	177	240	288	334	199
San Antonio, Tex.	302	282	237	191	141	122	134	168	203	218	261	292	213
Santa Maria, Calif.	329	297	254	203	151	122	127	167	233	267	307	336	233
Schenectady, N.Y.	215	193	145	106	62	50	63	97	133	165	200	217	137
Seattle, Wash.	242	209	150	84	44	29	34	60	118	174	216	228	132
Tucson, Ariz.	304	286	281	216	172	144	151	195	264	322	358	343	253
Washington, D.C.	267	190	196	145	75	64	101	124	153	182	215	247	163

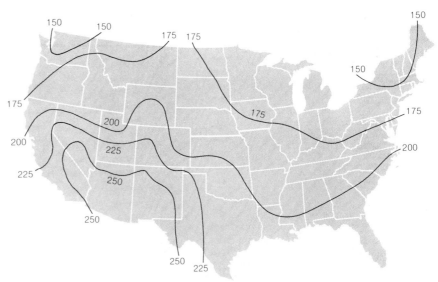

FIGURE F.1

Solar energy zones in the United States. The numbers give the annual average radiation on a horizontal surface, in watts per square meter (averaged over a full year and over all hours of the day and night). *Note*: solar radiation may vary significantly between nearby locations because of local variations in cloudiness. Some of these variations are not shown here, and caution should be used if this map is used to estimate available solar energy for a particular location.

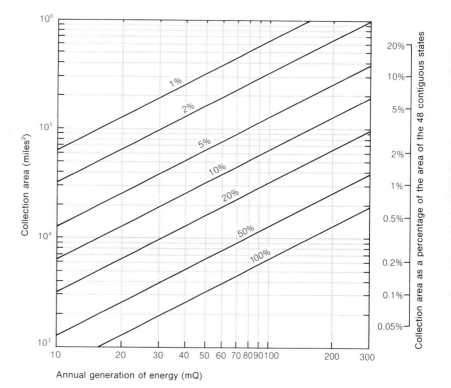

FIGURE F.2

Potentially available solar energy in the United States. The graph shows the required collection area versus the annual amount of energy derived from solar energy for various possible conversion efficiencies. (For example, at an efficiency of 2%, an area of 10^5 square miles could supply 30 mQ/yr.) This figure is based on an average value of 200 W/m² on a horizontal surface.

G

Degree-Days

FIGURE G.1
Degree-day zones in the United States (annual totals). *Note:* cities that are close together sometimes have significantly different degree-day totals. Some of these local variations are not shown here, and caution should be used if this map is used to estimate the degree-day total for a particular location.

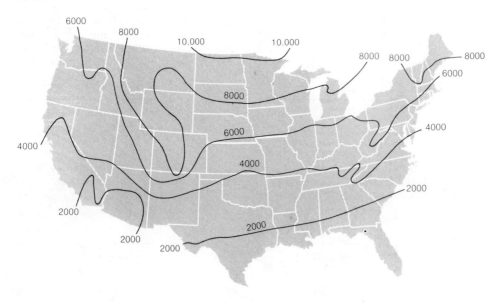

TABLE G.1   Degree-days of selected cities for a typical year.

	July	Aug.	Sept.	Oct.	Nov.	Dec.	Jan.	Feb.	Mar.	Apr.	May	June	Annual total
Atlanta, Ga.	0	0	18	127	414	626	639	529	437	168	25	0	2983
Barrow, Alaska	803	840	1035	1500	1971	2362	2517	2332	2468	1944	1445	957	20174
Bismarck, N. Dak.	34	28	222	577	1083	1463	1708	1442	1203	645	329	117	8851
Boise, Idaho	0	0	132	415	792	1017	1113	854	722	438	245	81	5809
Boston, Mass.	0	9	60	316	603	983	1088	972	846	513	208	36	5634
Caribou, Maine	78	115	336	682	1044	1535	1690	1470	1308	858	468	183	9767
Cleveland, Ohio	9	25	105	384	738	1088	1159	1047	918	552	260	66	6351
Columbia, Mo.	0	0	54	251	651	967	1076	874	716	324	121	12	5046
Dodge City, Kans.	0	0	33	251	666	939	1051	840	719	354	124	9	4986
El Paso, Tex.	0	0	0	84	414	648	685	445	319	105	0	0	2700
Fresno, Calif.	0	0	0	78	339	558	586	406	319	150	56	0	2492
Greensboro, N.C.	0	0	33	192	513	778	784	672	552	234	47	0	3805
Honolulu, Hawaii	0	0	0	0	0	0	0	0	0	0	0	0	0
Indianapolis, Ind.	0	0	90	316	723	1051	1113	949	809	432	177	39	5699
Little Rock, Ark.	0	0	9	127	465	716	756	577	434	126	9	0	3219
Miami, Fla.	0	0	0	0	0	40	56	36	9	0	0	0	141
New York, N.Y.	0	0	30	233	540	902	986	885	760	408	118	9	4871
Oklahoma City, Okla.	0	0	15	164	498	766	868	664	527	189	34	0	3725
Omaha, Nebr.	0	12	105	357	828	1175	1355	1126	939	465	208	42	6612
Rapid City, S. Dak.	22	12	165	481	897	1172	1333	1145	1051	615	326	126	7345
St. Cloud, Minn.	28	47	225	549	1065	1500	1702	1445	1221	666	326	105	8879
Salt Lake City, Utah	0	0	81	419	849	1082	1172	910	763	459	233	84	6052
San Antonio, Tex.	0	0	0	31	207	363	428	286	195	39	0	0	1549
Santa Maria, Calif.	99	93	96	146	270	391	459	370	363	282	233	165	2967
Schenectady, N.Y.	0	22	123	422	756	1159	1283	1131	970	543	211	30	6650
Seattle, Wash.	50	47	129	329	543	657	738	599	577	396	242	117	4424
Tucson, Ariz.	0	0	0	25	231	406	471	344	242	75	6	0	1800
Washington, D.C.	0	0	33	217	519	834	871	762	626	288	74	0	4224

H
Energy Content of Fuels

TABLE H.1 Energy content.[a]

Fuel	Values		
	(Commonly used units)	(BTU/ton)	(J/kg)
Coal (bituminous and anthracite)		25×10^6	29×10^6
Lignite		10×10^6	12×10^6
Peat		3.5×10^6	4×10^6
Crude oil	5.6×10^6 BTU/barrel	37×10^6	43×10^6
Gasoline	5.2×10^6 BTU/barrel	38×10^6	44×10^6
NGL's (Natural gas liquids)	4.2×10^6 BTU/barrel	37×10^6	43×10^6
Natural gas[b]	1030 BTU/ft³	47×10^6	55×10^6
Hydrogen gas[b]	333 BTU/ft³	107×10^6	124×10^6
Methanol (methyl alcohol)	6×10^4 BTU/gal	17×10^6	20×10^6
Charcoal		24×10^6	28×10^6
Wood	20×10^6 BTU/cord	12×10^6	14×10^6
Miscellaneous farm wastes		12×10^6	14×10^6
Dung		15×10^6	17×10^6
Assorted garbage and trash		10×10^6	12×10^6
Bread	1100 kcal/lb	9×10^6	10×10^6
Butter	3600 kcal/lb	29×10^6	33×10^6
Fission	200 MeV/fission	7×10^{13} [c]	8×10^{13} [c]
		5×10^{11} [d]	5.8×10^{11} [d]
D-D Fusion (deuterium)	7 MeV/deuteron	2.9×10^{14} [e]	3.3×10^{14} [e]
		8.6×10^{10} [f]	10^{11} [f]
D-T Fusion (lithium)[g]	7 MeV/Li nucleus	8.4×10^{13}	9.7×10^{13}
Complete "mass-energy conversion"[h]	931 MeV/amu	7.7×10^{16}	9×10^{16}

a These data are only intended for use in making _estimates_ of available energy. Various types of wood, for example, have energy values covering a rather wide range; different samples of coal, oil, and other fuels also have varying energy values. Various fuels obtained from the processing of oil (for example, residual oil, kerosene, various types of gasoline) have energy values per unit mass within about 20% of those listed for crude oil and gasoline.

b Quantities of natural gas are usually reported in cubic feet, the volume of gas at a pressure of 1 atmosphere and a temperature of 60°F, or in thousands of cubic feet (sometimes abbreviated Mcf).

c per ton or kilogram of nuclei undergoing fission.

d per ton or kilogram of uranium metal, when only the U^{235} (abundance 0.72%) is used.

e per ton or kilogram of pure deuterium.

f per ton or kilogram of hydrogen, containing 0.015% deuterium. (This is the natural abundance of deuterium; in a typical sample of hydrogen, 0.015% of the atoms are deuterium atoms, or 'heavy hydrogen'.)

g The data for D-T fusion are based on the assumption that deuterium is available in unlimited quantities, that tritium is 'bred' from lithium as discussed in Section J, and that energy production is limited by the availability of lithium.

h The data for complete mass-energy conversion are given for purposes of comparison only; no practical 'fuel' is known that would yield this much energy.

I

Fossil Fuels

TABLE I.1 Fossil fuel resources: estimates of eventual total production.
NOTE: Such estimates are subject to considerable uncertainty and should only be considered as giving the order of magnitude of the eventual total production. Estimates of production of energy from sources already in use are much more reliable than they are for tar sands and oil shales. The amount of oil that *exists* in the world's oil shales is estimated to be about 2×10^{15} barrels, 10,000 times larger than the figure given here; most of this oil is probably unobtainable, but the situation could be changed either by new technological developments or by changing economic conditions.

	United States		World (United States included)	
Fuel	Physical units	Approximate energy content (Q)	Physical units	Approximate energy content (Q)
Coal and lignite	1.6×10^{12} tons	37	8.4×10^{12} tons	170
Crude oil	200×10^9 barrels	1.1	2100×10^9 barrels	12
Natural gas liquids (NGL's)	40×10^9 barrels	0.17	400×10^9 barrels	1.7
Natural gas	1.1×10^{15} ft^3	1.1	12×10^{15} ft^3	12
Canadian tar sands	–	–	300×10^9 barrels	1.7
Oil shales	80×10^9 barrels	0.45	190×10^9 barrels	1.1
Total	–	40	–	200

TABLE I.2. Major fossil fuels: world and United States
consumption rates in relation to resources.

A. World (United States included)

	Coal and lignite	Petroleum liquids (crude oil and NGLs)	Natural gas
Resources (Table I.1)	8.4×10^{12} tons	2500×10^9 barrels	12×10^{15} ft^3
Cumulative production *	0.20×10^{12} tons (2.3% of resources)	530×10^9 barrels (21% of resources)	1.1×10^{15} ft^3 (9% of resources)
Production rate **	4.6×10^9 tons/yr	21×10^9 barrels/yr	53×10^{12} ft^3/yr
Time remaining ***	1800 years	95 years	210 years

B. United States

	Coal and lignite	Petroleum liquids (crude oil and NGLs)	Natural gas
Resources (Table I.1)	1.6×10^{12} tons	240×10^9 barrels	1.1×10^{15} ft^3
Cumulative production *	0.05×10^{12} tons (3% of resources)	154×10^9 barrels (64% of resources)	0.6×10^{15} ft^3 (55% of resources)
Production rate **	0.8×10^9 tons/yr	3.7×10^9 barrels/yr	16×10^{12} ft^3/yr
Time remaining ***	2000 years	23 years	31 years

    *   Cumulative production through 1983.
    ** Production rate as of 1983.
    *** Time remaining until resources would be exhausted,
        if production were to continue at the 1983 rate.

TABLE I.3. World and United States fossil fuel production rates, annual values, 1850-1983.

A. World (United States included).

	Coal and lignite (10^6 tons)	Crude oil (10^6 barrels)	NGLs (10^6 barrels)	Natural gas (10^9 ft^3)
1850	110			
1860	152	1		
1870	238	6		
1880	371	30		
1890	566	77		249
1900	851	149		251
1905	1041	215		374
1910	1284	328		540
1915	1315	432	2	665
1920	1488	689	9	835
1925	1512	1069	28	1240
1930	1558	1412	57	2002
1935	1452	1655	45	1973
1940	1854	2150	77	2792
1945	1516	2595	120	4263
1950	2000	3803	190	6683
1955	2345	5626	294	10239
1960	2870	7674	376	16054
1965	3034	11060	513	24130
1970	3244	16531	746	36195
1972	3271	18395	892	40992
1974	3511	20386	1020	43950
1976	3772	20976	1092	46120
1978	3921	21921	1169	48730
1980	4197	21791	1319	53780
1982	4382	19404	1377	53600
1983	4601	19303	1430	52700

B. United States.

	Coal and lignite (10^6 tons)	Crude oil (10^6 barrels)	NGLs (10^6 barrels)	Natural gas (10^9 ft^3)
1850	8			
1860	20	1		
1870	40	5		
1880	79	26		
1890	158	46		239
1900	270	63		236
1905	393	135		351
1910	502	210		509
1915	532	281	2	627
1920	658	443	9	801
1925	582	764	27	1176
1930	537	898	53	1913
1935	425	997	39	1920
1940	512	1353	70	2646
1945	633	1714	112	3902
1950	560	1974	182	6054
1955	491	2484	281	9053
1960	434	2575	340	12333
1965	527	2849	442	15379
1970	613	3517	606	21015
1972	602	3455	637	21580
1974	610	3203	616	20713
1976	685	2976	587	19098
1978	670	3178	572	19122
1980	830	3147	576	19403
1982	838	3157	566	17758
1983	785	3171	569	15972

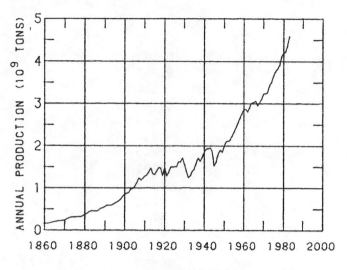

FIGURE I.1. WORLD PRODUCTION
OF COAL AND LIGNITE.

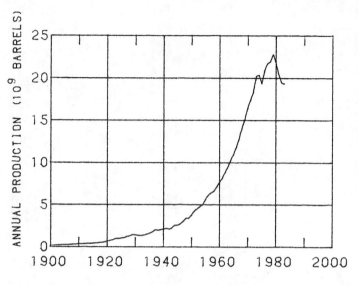

FIGURE I.2. WORLD
PRODUCTION OF CRUDE OIL.

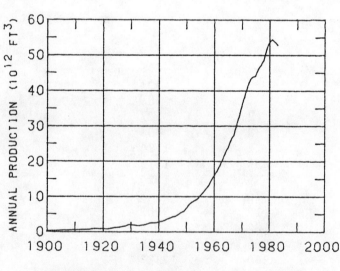

FIGURE I.3. WORLD
PRODUCTION OF NATURAL GAS.

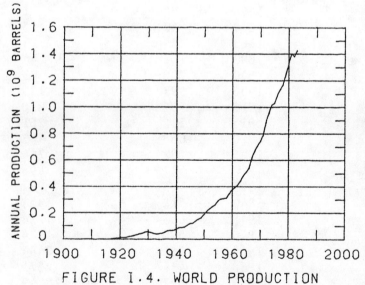

FIGURE I.4. WORLD PRODUCTION
OF NATURAL GAS LIQUIDS.

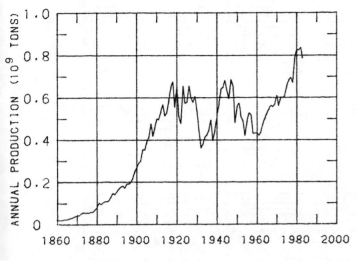

FIGURE I.5. UNITED STATES
PRODUCTION OF COAL AND LIGNITE.

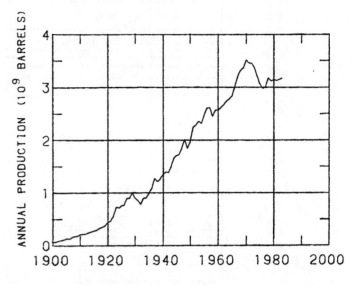

FIGURE I.6. UNITED STATES
PRODUCTION OF CRUDE OIL.

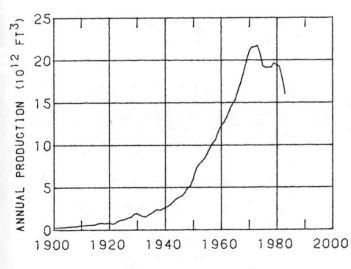

FIGURE I.7. UNITED STATES
PRODUCTION OF NATURAL GAS.

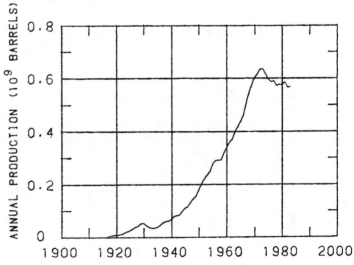

FIGURE I.8. UNITED STATES
PRODUCTION OF NATURAL GAS LIQUIDS.

41

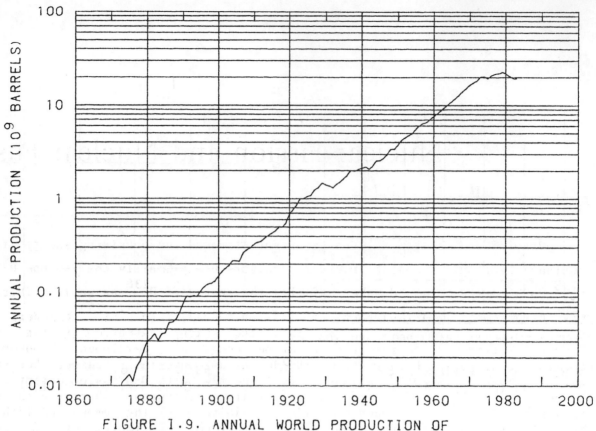

FIGURE I.9. ANNUAL WORLD PRODUCTION OF
CRUDE OIL. (THE SAME DATA AS IN FIGURE I.2,
ON A SEMILOGARITHMIC GRAPH.)

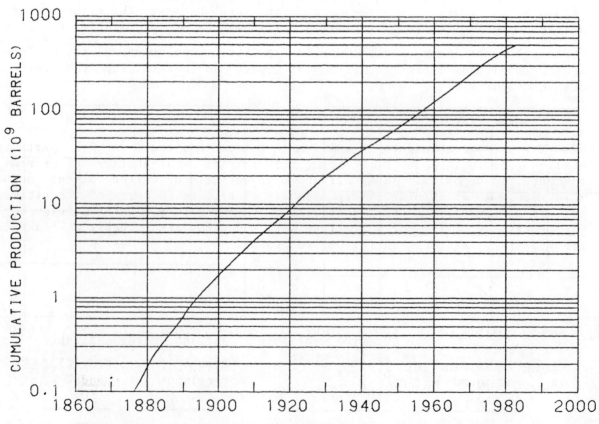

FIGURE I.10. CUMULATIVE PRODUCTION OF CRUDE OIL.
(FOR EXAMPLE, THE TOTAL PRODUCTION IN ALL
THE YEARS UP TO AND INCLUDING 1980 WAS
APPROXIMATELY 450 BILLION BARRELS.)

Nuclear Fission and Nuclear Fusion

Uranium and thorium are the important elements for obtaining energy from nuclear fission. Present nuclear reactors use primarily the isotope U^{235}. U^{238}-Pu^{239} breeders would make available the energy of U^{238} as well. Thorium may become an important source of fission energy if Th^{232}-U^{233} breeders are developed. Lithium will be an important energy resource if the D-T (deuterium-tritium) fusion process can be successfully developed. [Deuterium is also required for both D-T and D-D (deuterium-deuterium) fusion, but the amount of deuterium in the oceans is so large that with D-D fusion essentially no constraints will be imposed by availability of resources, whereas with D-T fusion, production of energy will be limited by the amount of lithium available and not by the supply of deuterium.] Estimates of available resources of uranium, thorium, and lithium are subject to great uncertainties, in part because until recently there has been little interest in these elements as energy resources, and in part because the amount of energy available per unit mass is so large that it may become profitable to extract these elements from very low-grade deposits. Development of the U^{238}-Pu^{239} breeder will have a particularly great impact in determining the value of uranium deposits.

Uranium resources

The average abundance of uranium in the earth's crust is 2.7 parts per million (Table E.2). More important is the availability of uranium in concentrated deposits. Table J.1 shows an estimate of United States uranium resources, in which uranium deposits are classified in terms of the estimated cost of mining, from a 1983 Department of Energy report. Uranium resources in the remainder of the world are less well known but are probably at least five times as large as those in the United States.

TABLE J.1. United States uranium resources.

Price (dollars per pound of ore)	Uranium available at this price or less (10^6 tons)	Fission energy available	
		From U^{235} only (Q)	From U^{235} and U^{238} (Q)
30	1.3	0.7	90
50	2.6	1.3	180
100	4.3	2.2	300

Thorium resources

The average abundance of thorium in the earth's crust is 9.6 parts per million (Table E.2). Rocks containing low-grade thorium deposits are found near the surface in many parts of the world. As one example, the Conway granites in New Hampshire, covering an area of about 750 km^2 and probably extending to a depth of several kilometers, contain 150 grams of thorium per cubic meter. The Conway granites down to a depth of 1 km thus contain approximately 1.2×10^8 tons of thorium, equivalent to a fission energy of about 8500 Q.

Lithium resources

The average abundance of lithium in the earth's crust is 20 parts per million (Table E.2). The extent of high-grade lithium deposits is less well known. According to one estimate, the presently known lithium resources of the United States amount to about 10^7 tons of lithium metal, equivalent to a fusion energy of about 800 Q.

Some Fusion Reactions

[p = proton (H^1), n = neutron, d = deuteron (H^2),
  t = tritium nucleus (H^3), α = alpha particle (He^4)]

$$d + d \rightarrow t + p + 4.0 \text{ MeV} \quad (1)$$

$$d + d \rightarrow He^3 + n + 3.3 \text{ MeV} \quad (2)$$

$$d + t \rightarrow \alpha + n + 17.6 \text{ MeV} \quad (3)$$

$$d + He^3 \rightarrow p + \alpha + 18.3 \text{ MeV} \quad (4)$$

$$Li^6 + n \rightarrow t + \alpha + 4.8 \text{ MeV} \quad (5)$$

D-D fusion. Reactions (1) and (2) occur with roughly equal probabilities. t and He^3 then react according to (3) and (4) so that the net result is:

$$6d \rightarrow 2p + 2n + 2\alpha + 43.2 \text{ MeV},$$

or about 7 MeV per deuteron.

D-T fusion (tritium breeding with lithim). Reactions (3) and (5) together yield:

$$d + Li^6 \rightarrow 2\alpha + 22.4 \text{ MeV}.$$

Other reactions also occur, some involving the more abundant isotope, Li^7. A reasonable expectation for the average energy release is about 7 MeV per lithium nucleus.

The History of Energy Production and Consumption in the World and the United States

Notes-

1. In these tables, the sum of individual values may not be exactly equal to the appropriate total because of independent rounding.

2. For the United States, account is taken of electrical generation from winds, geothermal and solar energy, and combustion of wastes [sources listed as 'other' in Section L], but not of other uses of these forms of energy, except in Figure K.6 where, for historical purposes, the use of fuel wood is included. This 'other' energy (about 0.14 mQ in 1983, 0.2% of total United States energy consumption) is included in the totals in Table K.2B but is not listed separately. In the electrical tables (K.3B and K.4B), it is arbitrarily included with 'fossil fuel'.

3. The United States imports some of its hydropower from Canada (about 33×10^9 kWh in 1983, 9% of our hydroelectricity and 1.4% of all our electrical energy); this energy is included in 'United States electrical generation' in Table K.4B.

TABLE K.1. Population of the world (United States included) and of the United States (in millions).

	World	United States
1900	1590	76
1905	1636	84
1910	1686	92
1915	1740	101
1920	1811	106
1925	1910	116
1930	2015	123
1935	2129	127
1940	2249	133
1945	2357	140
1950	2486	152
1955	2713	166
1960	2982	181
1965	3289	194
1970	3696	205
1972	3875	210
1974	4028	214
1976	4178	218
1978	4325	223
1980	4477	228
1982	4648	232
1983	4721	234

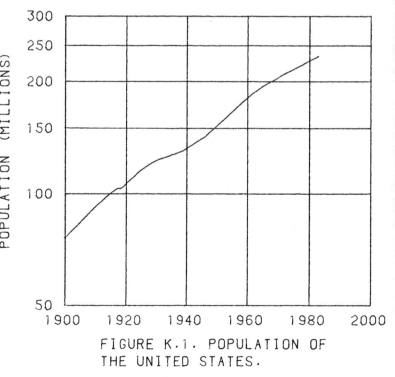

FIGURE K.1. POPULATION OF THE UNITED STATES.

TABLE K.2A. Annual world-wide energy consumption (United States included), 1900-1983.

	Coal & lignite (mQ)	Oil (mQ)	Natural gas (mQ)	Hydro-power (mQ)	Nuclear energy (mQ)	Total energy (mQ)	Average rate (GW)	Average per capita rate (W)
1900	20	0.8	0.3	0.3		21	713	448
1905	24	1.2	0.4	0.6		27	886	542
1910	30	1.8	0.6	1.1		33	1118	663
1915	31	2.4	0.7	1.8		35	1183	680
1920	34	3.9	0.9	2.1		41	1374	759
1925	34	6.1	1.3	2.1		44	1467	768
1930	35	8.1	2.1	2.9		48	1619	803
1935	33	9.5	2.0	2.9		47	1568	737
1940	41	12	2.9	3.7		60	1989	885
1945	34	15	4.4	4.4		58	1948	827
1950	44	22	6.9	5.1		78	2594	1044
1955	49	33	11	6.1		99	3295	1214
1960	61	45	17	7.9		130	4330	1452
1965	63	64	25	9.7	0.3	162	5406	1644
1970	67	96	37	12	0.9	213	7129	1929
1972	68	107	42	13	1.6	232	7740	1997
1974	72	118	45	15	2.6	254	8470	2103
1976	78	122	48	15	4.5	267	8915	2134
1978	81	128	50	17	6.5	282	9417	2177
1980	86	128	55	18	7.4	295	9853	2201
1982	90	114	55	19	9.5	288	9638	2074
1983	95	114	54	20	10	293	9783	2072

TABLE K.2B. Annual United States energy consumption, 1900-1983.

	Coal & lignite (mQ)	Oil (mQ)	Natural gas (mQ)	Hydro-power (mQ)	Nuclear energy (mQ)	Total energy (mQ)	Average rate (GW)	Average per capita rate (W)
1900	6.8	0.2	0.3	0.2		7.6	253	3319
1905	10	0.6	0.4	0.4		11	379	4524
1910	13	1.0	0.5	0.5		15	494	5341
1915	13	1.4	0.7	0.6		16	535	5320
1920	16	2.7	0.8	0.7		20	658	6185
1925	15	4.3	1.2	0.6		21	696	6011
1930	14	5.9	2.0	0.7		22	742	6031
1935	11	5.7	2.0	0.8		19	636	5002
1940	13	7.8	2.7	0.8		24	797	6013
1945	16	10	4.0	1.3		31	1049	7467
1950	13	13	6.2	1.4		34	1135	7453
1955	12	18	9.2	1.4		40	1330	8017
1960	10	20	13	1.6		45	1497	8284
1965	12	23	16	2.0		54	1796	9246
1970	13	30	22	2.6	0.2	67	2243	10939
1972	12	33	23	2.8	0.6	72	2392	11396
1974	13	33	22	3.3	1.3	72	2422	11324
1976	14	35	20	3.1	2.1	74	2482	11384
1978	14	38	20	3.1	3.0	78	2607	11714
1980	15	34	20	3.1	2.7	76	2536	11136
1982	15	30	19	3.6	3.1	71	2366	10196
1983	16	30	17	3.9	3.2	71	2360	10074

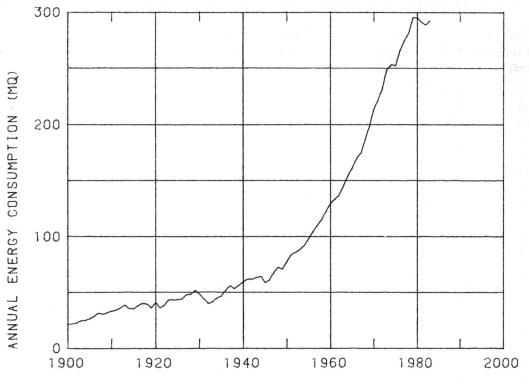

FIGURE K.2. ANNUAL WORLD CONSUMPTION OF ENERGY.

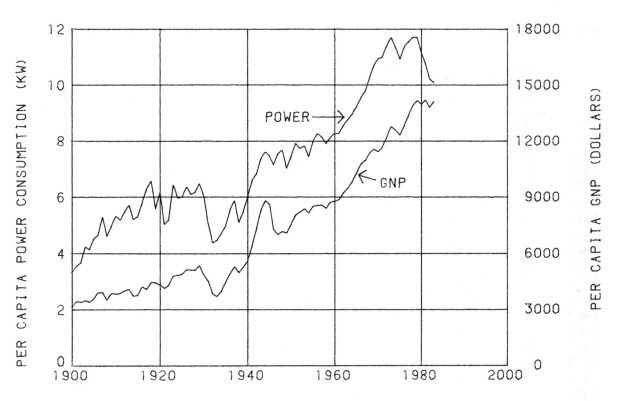

FIGURE K.3. PER CAPITA AVERAGE POWER CONSUMPTION AND PER
CAPITA GROSS NATIONAL PRODUCT IN THE UNITED STATES.
(GNP FIGURES ARE GIVEN IN 1983 DOLLARS; DOLLAR FIGURES FOR
PREVIOUS YEARS HAVE BEEN INCREASED TO TAKE ACCOUNT OF INFLATION.)

47

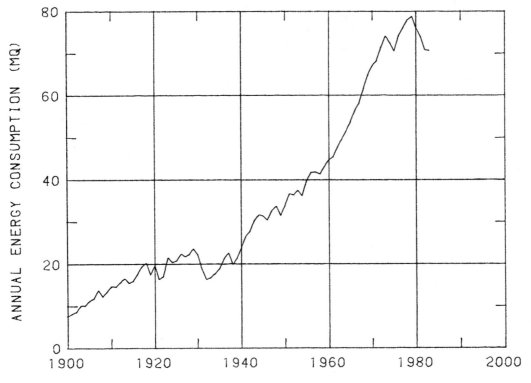

FIGURE K.4.   ANNUAL UNITED STATES CONSUMPTION OF ENERGY.

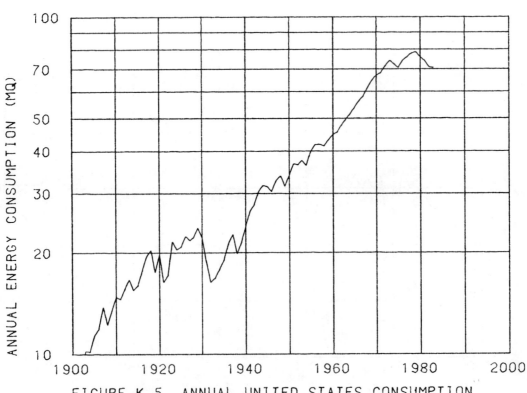

FIGURE K.5. ANNUAL UNITED STATES CONSUMPTION
OF ENERGY. (THE SAME DATA AS IN FIGURE K.4,
ON A SEMILOGARITHMIC GRAPH.)

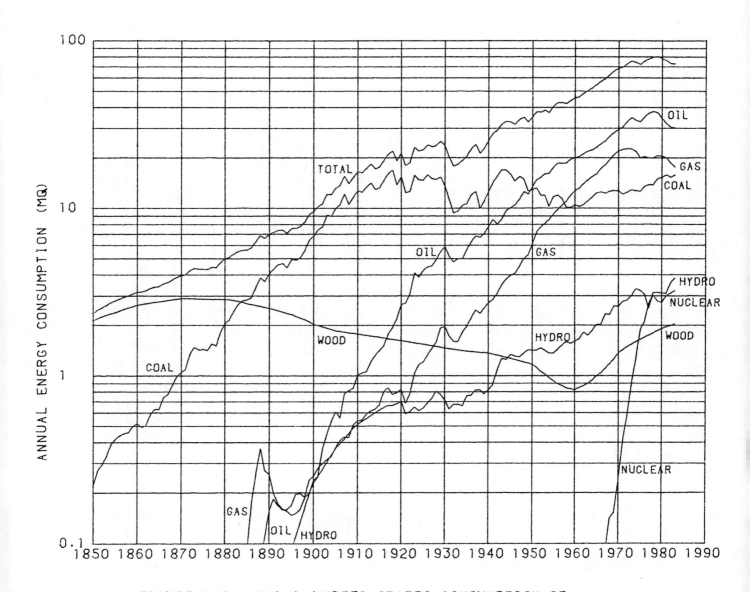

FIGURE K.6. ANNUAL UNITED STATES CONSUMPTION OF
ENERGY FROM VARIOUS SOURCES SINCE 1850.

ENERGY FROM FUEL WOOD IS INCLUDED HERE, BOTH AS
A SEPARATE ITEM AND AS PART OF THE TOTAL.
EXCEPT AS USED FOR ELECTRICAL GENERATION,
WOOD IS NOT INCLUDED IN OTHER GRAPHS AND TABLES.

49

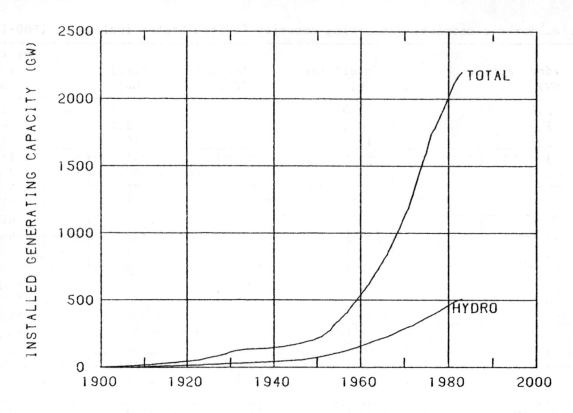

FIGURE K.7. INSTALLED ELECTRICAL
GENERATING CAPACITY, WORLD-WIDE.

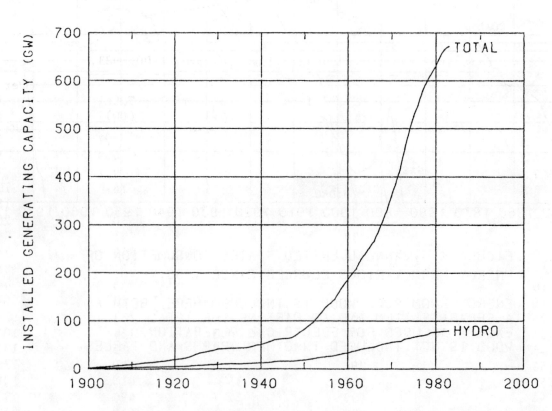

FIGURE K.8. INSTALLED ELECTRICAL GENERATING
CAPACITY OF THE UNITED STATES.

TABLE K.3A. World electrical generating capacity (United States included), 1900-1983.

	Hydroelectric (GW)	Fossil fuel (GW)	Nuclear (GW)	Total (GW)	Per capita total (W)
1900	1.3	1.6		2.9	1.8
1905	2.5	4.3		6.8	4.2
1910	5.1	11		16	9.7
1915	9.7	21		31	18
1920	14	30		45	25
1925	23	49		72	38
1930	30	85		115	57
1935	37	99		135	64
1940	45	103		147	66
1945	54	117		171	73
1950	76	141		218	88
1955	112	236		348	128
1960	160	381	0.9	542	182
1965	220	563	6.4	789	240
1970	291	816	19	1125	304
1972	313	948	35	1297	335
1974	353	1100	60	1513	376
1976	389	1261	87	1736	416
1978	424	1321	117	1862	431
1980	465	1406	142	2013	450
1982	501	1491	170	2161	465
1983	512	1517	176	2204	467

TABLE K.3B. United States electrical generating capacity, 1900-1983.

	Hydroelectric (GW)	Fossil fuel (GW)	Nuclear (GW)	Total (GW)	Per capita total (W)
1900	1.0	1.2		2.1	28
1905	1.6	3.3		4.9	58
1910	2.5	6.6		9.1	98
1915	3.4	10		14	134
1920	4.8	15		19	183
1925	7.2	23		30	260
1930	9.7	32		41	334
1935	10	33		43	340
1940	12	39		51	384
1945	16	47		63	448
1950	19	64		83	544
1955	26	105		131	789
1960	33	152	0.3	186	1028
1965	45	209	0.9	255	1310
1970	56	298	6.5	360	1757
1972	57	345	15	418	1989
1974	64	400	32	496	2317
1976	68	439	43	550	2524
1978	72	477	50	599	2690
1980	77	503	51	631	2771
1982	79	529	60	667	2874
1983	80	530	63	673	2873

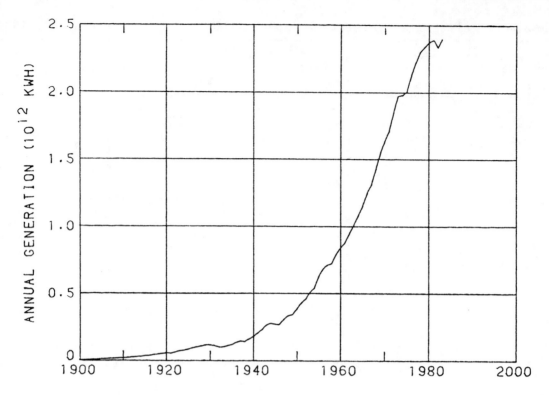

FIGURE K.9. GENERATION OF ELECTRICAL
ENERGY IN THE UNITED STATES.

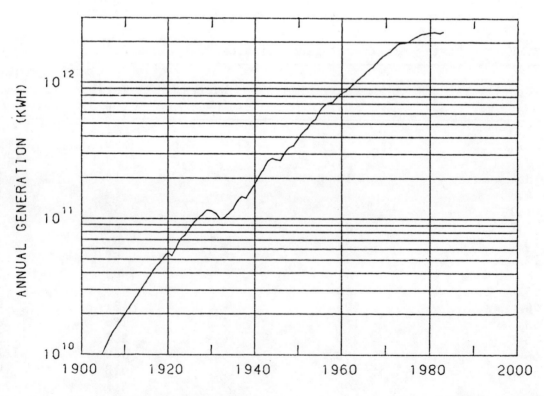

FIGURE K.10. GENERATION OF ELECTRICAL ENERGY
IN THE UNITED STATES. (THE SAME DATA AS IN
FIGURE K.9, ON A SEMILOGARITHMIC GRAPH.)

TABLE K.4A. World electrical generation, annual values (United States included), 1900-1983.

	Hydroelectric (10^9 kWh)	Fossil fuel (10^9 kWh)	Nuclear (10^9 kWh)	Total (10^9 kWh)	Per capita electrical energy per year (kWh)	Per capita average rate of use (W)
1900	3.7	1.3		5.0	3.1	0.36
1905	7.5	5.9		13	8.2	0.94
1910	17	16		33	20	2.2
1915	35	38		74	42	4.8
1920	54	68		122	67	7.7
1925	79	110		189	99	11
1930	127	173		300	149	17
1935	145	195		340	160	18
1940	199	302		500	222	25
1945	265	437		702	298	34
1950	339	607		946	380	43
1955	475	1065		1540	568	65
1960	688	1614	2.7	2304	773	88
1965	916	2437	24	3377	1027	117
1970	1172	3703	79	4953	1340	153
1972	1290	4269	148	5707	1473	168
1974	1434	4641	237	6312	1567	179
1976	1463	5195	411	7069	1692	193
1978	1629	5464	594	7688	1778	203
1980	1755	5791	681	8228	1838	210
1982	1833	5738	865	8436	1815	207
1983	1888	5682	917	8487	1798	205

TABLE K.4B. United States electrical generation, annual values, 1900-1983.

	Hydroelectric (10^9 kWh)	Fossil fuel (10^9 kWh)	Nuclear (10^9 kWh)	Total (10^9 kWh)	Per capita electrical energy per year (kWh)	Per capita average rate of use (W)
1900	2.8	1.2		4.0	52	5.9
1905	5.1	5.2		10	123	14
1910	8.6	11		20	214	24
1915	13	21		35	344	39
1920	19	38		57	531	61
1925	25	59		85	731	83
1930	36	79		115	931	106
1935	43	76		119	935	107
1940	50	130		180	1357	155
1945	85	187		271	1931	220
1950	101	288		389	2553	291
1955	116	513		629	3791	432
1960	150	694	0.5	844	4673	533
1965	195	960	3.4	1158	5958	680
1970	247	1370	22	1639	7993	912
1972	273	1526	54	1852	8825	1007
1974	317	1550	114	1981	9266	1057
1976	295	1647	191	2133	9784	1116
1978	303	1725	276	2305	10356	1181
1980	300	1824	251	2375	10432	1190
1982	343	1707	283	2333	10054	1147
1983	368	1739	294	2401	10251	1169

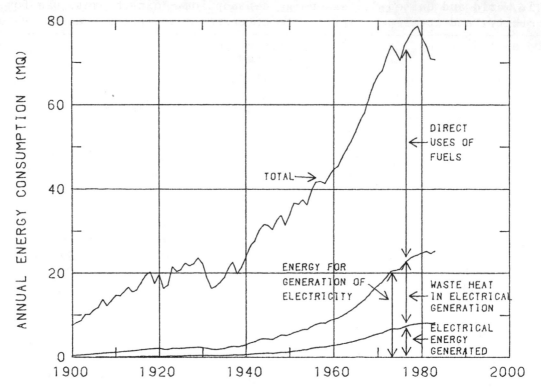

FIGURE K.11. TOTAL UNITED STATES ENERGY CONSUMPTION.
ALSO SHOWN IS THAT PORTION WHICH REPRESENTS
ELECTRICAL ENERGY GENERATED, AS WELL AS THE TOTAL
AMOUNT (WASTE HEAT INCLUDED) USED FOR THE
GENERATION OF ELECTRICITY.

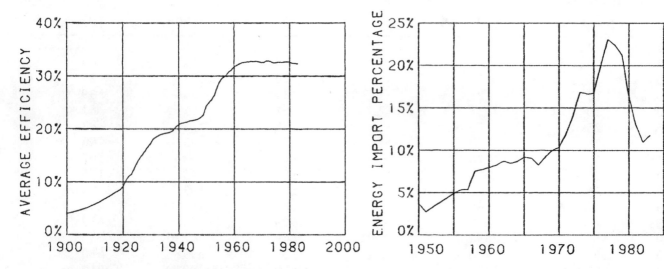

FIGURE K.12. AVERAGE EFFICIENCIES OF
AMERICAN STEAM-ELECTRIC
POWER PLANTS.

FIGURE K.13. UNITED STATES NET
IMPORTS OF ENERGY, AS A
PERCENTAGE OF TOTAL UNITED
STATES ENERGY CONSUMPTION.

TABLE K.5. World and United States energy consumption - direct uses, use for generation of electricity, and total per capita rate of energy consumption - annual values, 1900-1983.

A. WORLD (United States Included)

	Electrical Energy Generated		Energy Used for Generation of Electricity (Waste Heat Included)	Direct Uses of Energy	Total Energy Use	Average Per Capita Rate of Energy Use
	$(10^9$ kWh)	(mQ)	(mQ)	(mQ)	(mQ)	(W)
1900	5.0	0.017	0.45	21	21	448
1905	13	0.046	1.0	25	27	542
1910	33	0.11	2.1	31	33	663
1915	74	0.25	3.7	32	35	680
1920	122	0.42	4.8	36	41	759
1925	189	0.64	5.2	39	44	768
1930	300	1.0	6.8	42	48	803
1935	340	1.2	6.8	40	47	737
1940	500	1.7	9.2	50	60	885
1945	702	2.4	12	47	58	827
1950	946	3.2	14	63	78	1044
1955	1540	5.3	20	79	99	1214
1960	2304	7.9	26	103	130	1452
1965	3377	12	36	126	162	1644
1970	4953	17	52	161	213	1929
1972	5707	19	59	172	232	1997
1974	6312	22	66	187	254	2103
1976	7069	24	74	193	267	2134
1978	7688	26	80	202	282	2177
1980	8228	28	86	209	295	2201
1982	8436	29	89	200	288	2074
1983	8487	29	89	204	293	2072

B. UNITED STATES

	Electrical Energy Generated		Energy Used for Generation of Electricity (Waste Heat Included)	Direct Uses of Energy	Total Energy Use	Average Per Capita Rate of Energy Use
	$(10^9$ kWh)	(mQ)	(mQ)	(mQ)	(mQ)	(W)
1900	4.0	0.014	0.34	7.2	7.6	3319
1905	10	0.035	0.75	11	11	4524
1910	20	0.067	1.2	14	15	5341
1915	35	0.12	1.6	14	16	5320
1920	57	0.19	2.1	18	20	6185
1925	85	0.29	2.1	19	21	6011
1930	115	0.39	2.3	20	22	6031
1935	119	0.41	2.1	17	19	5002
1940	180	0.61	3.0	21	24	6013
1945	271	0.93	4.3	27	31	7467
1950	389	1.3	5.5	29	34	7453
1955	629	2.1	7.4	32	40	8017
1960	844	2.9	9.1	36	45	8284
1965	1158	4.0	12	42	54	9246
1970	1639	5.6	17	50	67	10939
1972	1852	6.3	19	52	72	11396
1974	1981	6.8	21	52	72	11324
1976	2133	7.3	22	52	74	11384
1978	2305	7.9	24	54	78	11714
1980	2375	8.1	25	51	76	11136
1982	2333	8.0	25	46	71	10196
1983	2401	8.2	25	45	71	10074

L
Sources and Uses of Energy
in the United States, 1983

Note - In these tables, the sum of individual values may not be exactly equal to
the appropriate total because of independent rounding.

TABLE L.1. Energy supply in the United States, 1983.

	Domestic supply (mQ)	Net imports (mQ)	Net energy supply (mQ)	(% of total)
Coal	18.0	-2.1	15.9	22.5%
Oil	20.9	9.2	30.1	42.6%
Natural Gas	16.6	0.9	17.5	24.7%
Total fossil fuels	55.4	8.0	63.4	89.8%
Hydropower	3.5	0.35	3.9	5.5%
Nuclear energy	3.2		3.2	4.6%
Other *	0.14		0.14	0.2%
Total	62.3 (88.2%)	8.3 (11.8%)	70.6	100 %

* Other: wind, geothermal and solar energy, and energy from combustion of
wood, refuse, etc., for the generaton of electrical energy. No attempt
is made to include energy from these sources that is used in ways other
than for electrical generation.

TABLE L.2. Use of energy in the United States, 1983.

	Input (mQ)	Direct uses (mQ)	Where it goes			Total (mQ)
			Electrical generation			
			Electrical energy (mQ)	Waste heat (mQ)	Total (mQ)	
Coal	15.9	2.7	4.3	8.9	13.2	15.9
Oil	30.1	28.5	0.54	1.0	1.6	30.1
Natural gas	17.5	14.1	1.0	2.3	3.3	17.5
Hydropower	3.9		1.3	2.6 *	3.9	3.9
Nuclear energy	3.2		1.0	2.2	3.2	3.2
Other	0.14		0.02	0.11	0.14	0.1
Total	70.6	45.3	8.2	17.1	25.3	70.6

* Hydroelectric 'waste heat' - As discussed in the Introduction, the
'efficiency' of hydroelectric generation is, for bookkeeeping purposes,
taken to be equal to the average efficiency of fossil-fuel generating
plants, with a corresponding amount of fictitious 'waste heat' being
created.

TABLE L.3. Direct uses of fuels, and use of energy for generation of electricity, 1983.

	Energy (mQ)	Percentage of national total
Direct uses of fossil fuels		
Coal	2.7	3.8%
Oil	28.5	40.4%
Natural gas	14.1	20.0%
Total direct uses	45.3	64.1%
Electrical generation		
Electrical energy generated	8.2	11.6%
Energy wasted in electrical generation	17.1	24.3%
Total for electrical generation	25.3	35.9%
Total energy consumption	70.6	100 %

TABLE L.4. Electrical generation, 1983.

	Electrical energy generated. (mQ)	Percentage of electrical energy generated	Energy used for generation (mQ)	Average efficiency
Coal	4.3	53.0%	13.2	32.9%
Oil	0.5	6.6%	1.6	34.3%
Natural gas	1.0	12.6%	3.3	30.8%
Hydroelectric power	1.3	15.3%	3.9	32.6% **
Nuclear power	1.0	12.2%	3.2	31.0%
Other	0.02	0.3%	0.14	16.3%
Total	8.2 *	100 %	25.3	32.3%

* Total amount of electrical energy generated: 8.2 mQ = 2.4×10^{12} kWh. Of the electrical energy generated, approximately 8% is lost in transmission. Most of the electrical energy is generated by public and private utilities, but some is generated by industries for internal use; a small amount is imported from Canadian hydroelectric plants (about 33×10^9 kWh, 9% of our hydroelectricity, and 1.4% of our total electrical energy).

** As discussed in the Introduction, the 'efficiency' of hydroelectric generation is, for bookkeeeping purposes, taken to be equal to the average efficiency of fossil-fuel generating plants, with a corresponding amount of fictitious 'waste heat' being created.

TABLE L.5. Use of fossil fuels, 1983.

	Coal (mQ)	Oil (mQ)	Natural gas (mQ)	Total (mQ)
Direct uses				
Residential & Commercial	0.3	2.4	7.2	9.9
Industrial	2.4	7.7	6.3	16.3
Transportation	0	18.4	0.6	19.0
Total direct uses	2.7	28.5	14.1	45.3
Use for generating electrical energy	13.2	1.6	3.3	18.1
Overall Total	15.9	30.1	17.5	63.4

TABLE L.6. Energy consumption by sector, 1983.

Sector	Direct uses of fuels (mQ)	Electricity (including waste heat from generation) (mQ)	Total energy consumption (mQ)	Percentage of national total
Residential & Commercial	9.9	15.8	25.7	36.4%
Industrial	16.3	9.5	25.9	36.6%
Transportation	19.0	0.03	19.0	27.0%
Total	45.3	25.3	70.6	100 %

TABLE L.7. Energy consumption within the residential
         and commercial sectors, percentages, 1973.

	Percentage of sector total	Percentage of national total
Space heating	50.9%	17.8%
Hot water	9.8%	3.4%
Air conditioning	9.2%	3.2%
Lighting (except street & highway lighting)	6.6%	2.3%
Non-energy uses *	6.1%	2.1%
Refrigeration	6.1%	2.1%
Cooking	3.4%	1.2%
Television	1.8%	0.6%
Home freezers	1.2%	0.4%
Home clothes drying	1.1%	0.4%
Miscellaneous home appliances	0.9%	0.3%
Street & highway lighting	0.4%	0.1%
Home dishwashers	0.3%	0.1%
Home washing machines	0.2%	0.1%
Other	2.0%	0.7%
Total	100 %	35.0%

* Largely use of petroleum products
  for highway paving, etc.

TABLE L.8. Energy consumption within the industrial sector, percentages, 1973.

	Percentage of sector total	Percentage of national total
Chemicals (production of basic chemicals, synthetic fibers, drugs, fertilizers, etc.)	20.8%	8.2%
Metals (smelting and refining, manufacture of basic metal products)	20.5%	8.1%
Petroleum refining, manufacture of paving and roofing materials, lubricants, etc.	11.5%	4.5%
Food manufacturing and processing	5.2%	2.1%
Paper products	5.2%	2.1%
Glass, concrete, asbestos, etc.	4.9%	1.9%
Other industries	31.9%	12.6%
Total	100 %	39.6%

TABLE L.9. Energy consumption within the transportation sector,
percentages, 1973.

	Percentage of sector total	Percentage of national total
Automobiles:		
local travel	36.0%	9.3%
intercity travel	19.1%	5.0%
	55.1%	14.3%
Trucks:		
local	16.9%	4.4%
intercity	6.2%	1.6%
	23.1%	6.0%
Planes:		
commercial passenger	4.1%	1.1%
military	2.8%	0.7%
air freight	1.0%	0.2%
private	0.5%	0.1%
	8.4%	2.1%
Railroads:		
freight	2.8%	0.7%
passenger	0.1%	0.03%
	2.9%	0.7%
Waterways	2.5%	0.6%
Pipelines	1.2%	0.3%
Buses:		
local travel	0.3%	0.1%
intercity travel	0.2%	0.05%
school buses	0.2%	0.04%
	0.7%	0.2%
Subways	0.2%	0.04%
Non-energy uses of fossil fuels (motor oils, greases, etc.)	0.8%	0.2%
Other transportation	5.5%	1.4%
Total	100 %	26.0%

M

Energy Requirements
for Electrical Appliances

TABLE M.1   Average wattage and estimated energy consumption of various appliances.

Appliance	Average power required (W)	Estimated electrical energy used per year (kWh)	Appliance	Average power required (W)	Estimated electrical energy used per year (kWh)
Air conditioner (window, 5000 BTU/hr)	1565	1390	Humidifier	175	163
			Iron	1000	144
Blanket	177	147	Microwave oven	1450	190
Broiler	1436	100	Radio	71	86
Carving knife	92	8	Radio-phonograph	110	110
Clock	2	17	Razor	14	2
Clothes dryer	4855	990	Refrigerator (12 ft^3)	240	730
Coffee maker	895	105	Refrigerator (12 ft^3, frostless)	320	1215
Deep-fat fryer	1450	83	Refrigerator-freezer (14 ft^3)	325	1135
Dishwasher	1200	363	Refrigerator-freezer (14 ft^3, frostless)	615	1830
Egg cooker	516	14			
Fan (attic)	370	290	Roaster	1333	205
Fan (circulating)	88	43	Sewing machine	75	11
Fan (window)	200	170	Stove	12,200	1175
Floor polisher	305	15	Sun lamp	280	16
Food blender	385	15	Television		
Food freezer (15 ft^3)	340	1200	Black & white, vacuum tubes	160	350
Food freezer (15 ft^3, frostless)	440	1760	Black & white, solid state	55	120
Food mixer	127	13	Color, vacuum tubes	300	660
Food waste disposer	445	30	Color, solid state	200	440
Frying pan	1200	185	Toaster	1145	40
Grill (sandwich)	1160	33	Tooth brush	7	0.5
Hair dryer	380	14	Trash compactor	400	50
Heat lamp	250	13	Vacuum cleaner	630	46
Heater (portable)	1320	175	Vibrator	40	2
Heating pad	65	10	Waffle iron	1115	22
Hot plate	1260	90	Washing machine (automatic)	512	103

Source: Electrical Energy Association, New York, N.Y.

Energy Requirements for Passenger and Freight Transportation

TABLE N.1   Energy requirements for passenger transportation.

Mode of transport	Maximum capacity (no. of passengers)*	Vehicle mileage (miles/gal)†	Passenger mileage (passenger-miles/gal)†	Energy consumption (BTU/passenger-mile)
Bicycle	1	1560	1560	80
Walking	1	470	470	260
Intercity bus	45	5	225	550
Commuter train (10 cars)	800	0.2	160	775
Subway train (10 cars)	1000	0.15	150	825
Volkswagen sedan**	4	30	120	1030
Local bus	35	3	105	1180
Intercity train (4 coaches)	200	0.4	80	1550
Motorcycle	1	60	60	2060
Automobile††	4	12	48	2580
747 jet plane	360	0.1	36	3440
727 jet plane	90	0.4	36	3440
707 jet plane	125	0.25	31	3960
United States SST (proposed)	250	0.1	25	4950
Light plane (2 seat)	2	12	24	5160
Executive jet plane	8	2	16	7740
Concorde SST	110	0.12	13	9400
Snowmobile	1	12	12	10,300
Ocean liner	2000	0.005	10	12,400

*The relative effectiveness of various modes of transportation can be drastically altered if a smaller number of passengers is carried.
†Miles per gallon of gasoline or the equivalent in food or in other fuel; all values must be regarded as approximate.
**Long-distance intercity travel.
††Typical American automobile, used partly for local travel and partly for long-distance driving.

TABLE N.2   Energy requirements for freight transportation.

Mode of transport	Mileage (ton-miles/gal)	Energy consumption (BTU/ton-mile)
Oil pipelines	275	450
Railroads	185	670
Waterways	182	680
Truck	44	2800
Airplane	3	42000

O
Exponential Growth

TABLE O.1   Doubling times for various rates of exponential growth.

Multiplication factor in each unit of time	Percentage increase per unit time	Doubling time
1.0	0	Infinite
1.01	1	69.7
1.02	2	35.0
1.03	3	23.4
1.04	4	17.7
1.05	5	14.2
1.06	6	11.9
1.07	7	10.2
1.08	8	9.0
1.09	9	8.0
1.10	10	7.3
1.12	12	6.1
1.14	14	5.3
1.16	16	4.7
1.18	18	4.2
1.20	20	3.8
1.25	25	3.1
1.30	30	2.6
1.40	40	2.1
1.50	50	1.7
1.75	75	1.2
2.00	100	1.0

Consumer Prices of Common Sources of Energy

The following table lists the average residential prices of electricity, natural gas, and fuel oil, and the retail price of various grades of gasoline (full-serve, taxes included). The overall consumer price index is given for comparison. Consumer prices vary widely from one part of the United States to another, and prices paid by industrial and commercial users are often much lower than those paid by individuals.

TABLE P.1. Prices of energy sources, 1960-1983.

	Electricity (1 kWh) (cents)	Natural gas (1000 ft^3) (dollars)	Fuel oil (1 gal) (dollars)	Gasoline Regular leaded (1 gal) (dollars)	Gasoline Regular unleaded (1 gal) (dollars)	Gasoline Premium unleaded (1 gal) (dollars)	Consumer price index (1967=100)
1960	2.47	0.97	0.15	0.31			88.7
1961	2.45	0.98	0.16	0.31			89.6
1962	2.41	0.98	0.16	0.31			90.6
1963	2.37	0.98	0.16	0.30			91.7
1964	2.31	0.98	0.16	0.30			92.9
1965	2.25	0.98	0.16	0.31			94.5
1966	2.20	0.99	0.16	0.32			97.2
1967	2.17	0.99	0.17	0.34			100.0
1968	2.14	1.00	0.17	0.34			104.2
1969	2.13	1.02	0.18	0.35			109.8
1970	2.18	1.07	0.19	0.36			116.3
1971	2.32	1.15	0.20	0.36			121.3
1972	2.42	1.21	0.20	0.37			125.3
1973	2.54	1.29	0.23	0.39			133.1
1974	3.10	1.43	0.35	0.53			147.7
1975	3.51	1.71	0.38	0.57			161.2
1976	3.73	1.98	0.41	0.59	0.61		170.5
1977	4.05	2.35	0.46	0.62	0.66		181.5
1978	4.31	2.56	0.49	0.63	0.67		195.4
1979	4.64	2.98	0.70	0.86	0.90		217.4
1980	5.36	3.68	0.97	1.19	1.25		246.8
1981	6.20	4.29	1.19	1.31	1.38	1.47	272.4
1982	6.86	5.17	1.16	1.22	1.30	1.41	289.1
1983	7.18	5.99	1.08	1.16	1.24	1.38	298.4

Q

Comparing Capital and Operating Costs.
The Cost of Borrowing Money

In assessing the cost of any proposal requiring a significant capital investment, it is important to be able to convert a capital cost into an equivalent annual cost, a process often referred to as amortizing the capital cost. This procedure is essential if one is trying to compare the cost of one method of producing energy, in which the capital investment is large and the fuel costs low, with a second method for which the reverse is true. If money could be borrowed with no interest charges, then, for instance, a capital investment of $1000 in equipment with a useful lifetime of ten years could be financed by paying the lender in ten equal installments of $100 each year. Because interest must be paid on the outstanding balance, the annual cost will be more than expected from this simple calculation, the excess amount depending on the annual interest rate and the length of time over which the loan is to be repaid.

Consider the situation in which the borrower agrees to pay the lender an equal sum of money each year (part of each payment as interest on the outstanding balance and part to reduce the size of the balance). We need to know the size of each annual payment that will produce the desired result; that is, that the final payment shall neatly reduce the balance to zero. The general formula is somewhat complicated, but there exist tables to facilitate the calculation. Table Q.1 is an example of such an amortization table. It shows, for an initial loan of $1: (1) the amount of each annual payment, and (2) the total amount that must be repaid (the amount of each annual payment multiplied by the number of years). Figures are given for several possible rates of interest and time periods. For actual loans, the figures shown in this table should be multiplied by the dollar amount of the loan.

With an interest rate of 7% and a time period of 20 years, for example, an initial loan of $1.00 can be paid back in 20 equal installments of $0.0944 (9.44 cents) each, a total payment of $1.89. With these figures, a capital investment of $1 is equivalent to an annual cost of $0.0944 per year, a capital investment of $1000 is equivalent to an annual cost of $94.4 per year, and so on. Observe that the total amount paid to the lender is 1.89 times as large as the amount borrowed. (For longer loans at higher interest rates, the total amount paid to the lender can be much greater than the amount originally borrowed, ten times as large, for example, for a 50-year loan at 20% interest.)

A useful approximate formula, one which is best for loans of long duration and relatively low interest, is given by the following equation: *

$$P = \frac{Ar}{1 - e^{-rN}} \qquad (1)$$

where A is the dollar amount of the loan, r is the interest rate (e.g., 0.07 for a 7% loan), N is the number of years over which the loan is to be repaid, and P is the required annual payment. (Here e is the base of natural logarithms, 2.71828 ...) As an example, if A = $1, r = 0.03 and N = 50, Eq. (1) gives P = 0.0386, the same as Table Q.1 to within 1%, and even for r = 0.15 and N = 10, Eq. (1) gives P = 0.193, the same as the exact result to within 3%. Eq. (1) can readily be used to calculate required payments for time periods and rates of interest which are not included in Table Q.1.

Whether one uses a table or an equation, inflation is a complicating factor in any such calculation. Suppose, for instance, that the average rate of inflation over the period of the loan will be 6%. Then if someone is willing to lend you money at 6% annual interest, what you really have, in effect, is an interest-free loan. One can make allowance for other numerical values in similar fashion, by subtracting the annual rate of inflation from the interest rate. For instance, if the annual rate of inflation is 10%, a 15% loan should be treated as if it were simply a 5% loan, for purposes of comparing capital and operating costs.

* Eq. (1) would be exactly correct if repayment were made <u>continuously</u> instead of in discrete annual installments. That is why Eq. (1) is more nearly correct for loans such as a 50-year loan, for which a large number of payments are made even though payments are only made once a year.

TABLE Q.1.   Annual and total cost of a $1 loan.

Annual interest rate	Number of years over which loan is repaid									
	10		20		30		40		50	
	Payments		Payments		Payments		Payments		Payments	
	Annual ($)	Total ($)	Annual ($)	Total ($)	Annual ($)	Total ($)	Annual ($)	Total ($)	Annual ($)	Total ($)
1%	0.106	1.06	0.0554	1.11	0.0387	1.16	0.0305	1.22	0.0255	1.28
2%	0.111	1.11	0.0612	1.22	0.0446	1.34	0.0366	1.46	0.0318	1.59
3%	0.117	1.17	0.0672	1.34	0.0510	1.53	0.0433	1.73	0.0389	1.94
4%	0.123	1.23	0.0736	1.47	0.0578	1.73	0.0505	2.02	0.0466	2.33
5%	0.130	1.30	0.0802	1.60	0.0651	1.95	0.0583	2.33	0.0548	2.74
6%	0.136	1.36	0.0872	1.74	0.0726	2.18	0.0665	2.66	0.0634	3.17
7%	0.142	1.42	0.0944	1.89	0.0806	2.42	0.0750	3.00	0.0725	3.62
8%	0.149	1.49	0.102	2.04	0.0888	2.66	0.0839	3.35	0.0817	4.09
9%	0.156	1.56	0.110	2.19	0.0973	2.92	0.0930	3.72	0.0912	4.56
10%	0.163	1.63	0.117	2.35	0.106	3.18	0.102	4.09	0.101	5.04
11%	0.170	1.70	0.126	2.51	0.115	3.45	0.112	4.47	0.111	5.53
12%	0.177	1.77	0.134	2.68	0.124	3.72	0.121	4.85	0.120	6.02
13%	0.184	1.84	0.142	2.85	0.133	4.00	0.131	5.24	0.130	6.51
14%	0.192	1.92	0.151	3.02	0.143	4.28	0.141	5.63	0.140	7.01
15%	0.199	1.99	0.160	3.20	0.152	4.57	0.151	6.02	0.150	7.51
16%	0.207	2.07	0.169	3.37	0.162	4.86	0.160	6.42	0.160	8.00
17%	0.215	2.15	0.178	3.55	0.172	5.15	0.170	6.81	0.170	8.50
18%	0.223	2.23	0.187	3.74	0.181	5.44	0.180	7.21	0.180	9.00
19%	0.230	2.30	0.196	3.92	0.191	5.73	0.190	7.61	0.190	9.50
20%	0.239	2.39	0.205	4.11	0.201	6.03	0.200	8.01	0.200	10.00

R
Radiation Exposure in the United States

TABLE R.1 Estimated exposure to ionizing radiation in the United States (1970).

Radiation source	Average dose rate per person (mrem/yr)
Natural sources	
Cosmic rays at ground level*	44
Rocks, soil and building materials†	40
Sources within the body (largely K⁴⁰)	18
Subtotal	102
Artificial sources	
Fallout from nuclear weapons testing	4
Medical and dental diagnosis and treatment	73
Nuclear power installations**	0.003
Occupational exposure	0.8
Miscellaneous††	2
Subtotal	80
Total	182

*Range within the United States: 38–75. See Table R.2 for variation with altitude.

†Range 15–140 depending on type of soil and building material.

**Nuclear power reactors and fuel reprocessing plants; estimated to increase to 0.4 mrem/yr by the year 2000.

††Television sets, airplane travel, etc.

TABLE R.2 Variation of cosmic ray dose rate with altitude above sea level.

Altitude (ft)	Dose rate* (mrem/yr)
0	40
2500	52
5000	68
7500	100
10,000	190
15,000	460
20,000	1100
30,000	4000
45,000	9000
55,000–100,000	13,000

*Values are approximate and vary with latitude and (at high altitudes) vary during the 11-year cycle of solar sun-spot activity. During occasional solar flares, dose rates at jet airplane altitudes of 10^6 mrem/yr may be observed for periods of a few hours. Dose rates in the Van Allen radiation belts (altitude 1000 miles and higher) may be as large as 10^7 mrem/yr.